Quantum Robotics

A Primer on Current Science and Future Perspectives

Synthesis Lectures on Quantum Computing

Editors
Marco Lanzagorta, *U.S. Naval Research Labs*
Jeffrey Uhlmann, *University of Missouri-Columbia*

Quantum Computer Science
Marco Lanzagorta and Jeffrey Uhlmann
2008

Quantum Walks for Computer Scientists
Salvador Elías Venegas-Andraca
2008

Quantum Robotics: A Primer on Current Science and Future Perspectives

Prateek Tandon, Stanley Lam, Ben Shih, Tanay Mehta, Alex Mitev, and Zhiyang Ong

ISBN: 978-3-031-01392-8   paperback
ISBN: 978-3-031-02520-4   ebook
ISBN: 978-3-031-03648-4   epub

DOI 10.1007/978-3-031-02520-4

A Publication in the Springer series
SYNTHESIS LECTURES ON QUANTUM COMPUTING

Lecture #10
Series Editors: Marco Lanzagorta, *U.S. Naval Research Labs*
                Jeffrey Uhlmann, *University of Missouri–Columbia*
Series ISSN
Print 1945-9726   Electronic 1945-9734

Quantum Robotics

A Primer on Current Science and Future Perspectives

Prateek Tandon
Stanley Lam
Ben Shih
Tanay Mehta
Alex Mitev
Zhiyang Ong
Quantum Robotics Group

SYNTHESIS LECTURES ON QUANTUM COMPUTING #10

ABSTRACT

Quantum robotics is an emerging engineering and scientific research discipline that explores the application of quantum mechanics, quantum computing, quantum algorithms, and related fields to robotics. This work broadly surveys advances in our scientific understanding and engineering of quantum mechanisms and how these developments are expected to impact the technical capability for robots to sense, plan, learn, and act in a dynamic environment. It also discusses the new technological potential that quantum approaches may unlock for sensing and control, especially for exploring and manipulating quantum-scale environments. Finally, the work surveys the state of the art in current implementations, along with their benefits and limitations, and provides a roadmap for the future.

KEYWORDS

Quantum Robotics, Quantum Computing, Quantum Algorithms

Contents

Preface

The Quantum Robotics Group was founded in March 2015. The group met every weekend over the course of a year to discuss different emerging topics related to quantum robotics. This book is the product of our lecture series.

Computational speedups in planning for complex environments, faster learning algorithms, memory and power efficiency using qubit representation of data, and capability to manipulate quantum phenomena are only some of the many exciting possibilities in the emerging world of quantum robotics. Robotic systems are likely to benefit from quantum approaches in many ways.

Our book serves as a roadmap for the emerging field of quantum robotics, summarizing key recent advances in quantum science and engineering and discussing how these may be beneficial to robotics. We provide both a survey of the underlying theory (of quantum computing and quantum algorithms) as well as an overview of current experimental implementations being developed by academic and commercial research groups. Our aim is to provide a starting point for readers entering the world of quantum robotics and a guide for further exploration in sub-fields of interest. From reading our exposition, we hope that a better collective understanding of quantum robotics will emerge.

Chapter 1 introduces our work and framework. In Chapter 2, we provide background on relevant concepts in quantum mechanics and quantum computing that may be useful for quantum robotics. From there, the survey delves into key concepts in quantum search algorithms (Chapter 3) that are built on top of the quantum computing primitives. Speedups (and other algorithmic advantages) resulting from the quantum world are also investigated in the context of robot planning (Chapter 4), machine learning (Chapter 5), and robot controls and perception (Chapter 6). Our book also highlights some of the current implementations of quantum engineering mechanisms (Chapter 7) as well as current limitations. Finally, we conclude with a holistic summary of potential benefits to robotics from quantum mechanisms (Chapter 8).

We hope you enjoy this work and, from it, are inspired to delve more into the exciting emerging world of quantum robotics.

Prateek Tandon, Stanley Lam, Ben Shih, Tanay Mehta, Alex Mitev, and Zhiyang Ong
January 2017

Acknowledgments

We would like to acknowledge Dr. Steven Adachi, Professor Marco Lanzagorta, Professor Jeffrey Uhlmann, and Dr. Henry Yuen, among others, for their helpful comments in the development of this work.

Notation

In this section, we detail some of the notation and conventions used throughout the book. In writing our work, we have attempted to use the same notation as the original cited publications to maintain a high fidelity to the original literature. However, in some cases, we have modified the notation to make equations easier to read.

STANDARD NOTATION

- $\mathbb{N}$ denotes the set of nonnegative integers.
- $\mathbb{R}$ denotes the set of real numbers.
- $\mathbb{C}$ denotes the space of complex numbers.
- $\mathbb{C}^N$ denotes the N-dimensional space of complex numbered vectors.

COMPUTER SCIENCE

We make extensive use of asymptotic notation. For two functions $f, g : \mathbb{N} \to \mathbb{N}$, we write:

- $f(x) = O(g(x))$ if and only if there exists a positive constant C and an integer x_0 such that $|f(x)| \leq C|g(x)|$ for all $x \geq x_0$.
- $f(x) = \Omega(g(x))$ if and only if $g(x) = O(f(x))$.
- $f(x) = \Theta(g(x))$ if and only if $f(x) = O(g(x))$ and $f(x) = \Omega(g(x))$.
- $a \oplus b$ refers to the XOR operation between two binary bits, a and b.

CALCULUS

- $\dot{f}(x)$ generally refers to the first derivative of the differentiable function $f(x)$.
- $\ddot{f}(x)$ generally refers to the second derivative of the differentiable function $f(x)$.

LINEAR ALGEBRA

- I generally refers to the identity matrix of appropriate size (unless otherwise stated).
- $\det(A)$ refers to the determinant of the matrix A.
- $A^\dagger$ refers to the conjugate transpose of A.
- F^+ denotes the Moore-Penrose pseudoinverse of F.

QUANTUM MECHANICS

- $\hbar$ refers to Planck's constant.

- $|\psi\rangle$ refers to a ket, which is generally a state vector for a quantum state.

- $\langle\psi|$ refers to a bra, the conjugate transpose of the vector $|\psi\rangle$.

- ρ generally refers to the density matrix of a quantum system (unless otherwise stated).

- U generally refers to a unitary matrix where $U^\dagger U = UU^\dagger = I$ (unless otherwise stated).

- $\langle\phi|\psi\rangle$ refers to the inner product between the vectors $|\phi\rangle$ and $|\psi\rangle$.

- $\langle\phi|A|\psi\rangle$ refers to the inner product between ϕ and $A\psi$.

- $|\phi\rangle \otimes |\psi\rangle$ refers to a tensor product between $|\phi\rangle$ and $|\psi\rangle$.

- $|\phi\rangle |\psi\rangle$ also refers to the tensor product between $|\phi\rangle$ and $|\psi\rangle$.

- $|\psi\rangle^{\otimes N}$ refers to the quantum state (in superposition) of the composite system with N interacting quantum systems, each having quantum state $|\psi\rangle$.

- σ_x often refers to the Pauli-X matrix $\begin{bmatrix} 0 & 1 \\ 1 & 0 \end{bmatrix}$.

- σ_y often refers to the Pauli-Y matrix $\begin{bmatrix} 0 & -i \\ i & 0 \end{bmatrix}$ where $i = \sqrt{-1}$.

- σ_z often refers to the Pauli-Z matrix $\begin{bmatrix} 1 & 0 \\ 0 & -1 \end{bmatrix}$.

- $[A, B]$ often refers to a commutation operator $[A, B] = AB - BA$.

- A set of matrices $\{K_i\}$ is a set of Kraus matrices if it satisfies $\sum_{i=1}^{\kappa} K_i^\dagger K_i = I_d$

CHAPTER 1

Introduction

A robot is a physical hardware embodied agent, situated and operating in an uncertain and dynamic real-world environment [Matarić, 2007]. Typical robots have sensors by which they can perceive their environment's state (as well as their own), manipulators for acting in and affecting their environment, electronic hardware capable of real-time computation and control, and sophisticated software algorithms.

The software algorithms are the "brains" of the robot, providing the principles for sensing, planning, acting, and learning with respect to the environment. These algorithms enable the robot to represent the joint robot-environment state and reason over sensor uncertainties and environment dynamics. A key hurdle to the development of more intelligent robotics has traditionally been computational tractability and scalability of algorithms. Robotic planning quickly becomes computationally infeasible for classical implementations as the time horizon for which an optimal plan must be formulated is increased. Classical robotic learning suffers from the curse of dimensionality. As dimensionality of sensor percept data increases and the hypothesis space over which it is interpreted becomes large, there exist fewer and fewer algorithms that can operate well to make sense of the sensor data while still being efficient.

The technological capabilities of classical robots are thus often pillared on fundamental development in systems and algorithms. Advances in sub-fields of robotics such as perception, planning, machine learning, and control push the intelligence periphery of what robots can do. The field of quantum robotics explores the applications of quantum principles to enhance software, hardware, and algorithmic capability in these areas.

1.1 WHAT DOES QUANTUM ROBOTICS STUDY?

Quantum robotics explores the application of the principles of quantum mechanics, quantum computing, quantum algorithms, and related fields to robotics. The quantum world is expected to provide many possible benefits to robot hardware and software intelligence capability.

Quantum computing theory predicts significant asymptotic speed ups in the worst-case time complexity for many classical algorithms used by robots to solve computational problems. Techniques such as quantum parallelism, Grover's algorithm, and quantum adiabatic optimization may improve asymptotic performance on classically NP-complete computational problems for robots.

Qubit ("quantum bit") representation of data is thought to be more scalable and power efficient than traditional binary bit representation of data. This may allow for gains in the processing of large amounts of data by robotic systems. While there are key limitations with storing and extracting data from a quantum memory, there are expected benefits even with the fundamental limitations. Whether the benefits are mostly for model building in offline mode or extend to real-time operation remains to be seen, but the potential for impact is surely there. In addition, the potential energy efficiency of quantum-scale circuitry and qubit hardware may bring down the power consumed by robotic systems.

Aside from providing potential computational software and hardware advantages for robots operating in classical environments, quantum approaches unlock new possibilities for robot sensing and control in environments governed by quantum dynamics. Quantum mechanical principles may be useful in engineering new quantum sensors and creating new quantum robot controllers that can operate on matter at a quantum scale. Many of the classical filtering algorithms (such as Kalman Filters or Hidden Markov Models) have quantum analogues and expected improvements in dealing with uncertainty, representational power, and with operating in quantum environments.

Quantum robotics is as much about science as it is engineering, and the emphasis of our field is on plausible science. Most quantum algorithms have highly specific conditions under which they work. Recognizing the rigorous scientific limitations of quantum methods is important for appropriate application in robotics.

1.2   AIM AND OVERVIEW OF OUR WORK

Our book serves as a roadmap for the emerging field of quantum robotics, summarizing key recent advances in quantum science and engineering and discussing how these may be beneficial to robotics. We provide both a survey of the underlying theory (of quantum computing and quantum algorithms) as well as an overview of current experimental implementations being developed by academic and commercial research groups. Our aim is to provide a starting point for readers entering the world of quantum robotics and a guide for further exploration in sub-fields of interest. From reading our exposition, we hope that a better collective understanding of quantum robotics will emerge.

In general, our work is written for an audience familiar with robotic algorithms. While our book provides brief introductions to classical methods commonly used in robotic planning, learning, sensing, and control, the reader may wish to brush up on the prerequisites from other readily available robotic textbooks. Our work does not, however, presume any prior knowledge of quantum mechanics or quantum computing.

In Chapter 2, we provide background on relevant concepts in quantum mechanics and quantum computing that may be useful for quantum robotics. From there, the survey delves into key concepts in quantum search algorithms (Chapter 3) that are built on top of the quantum computing primitives. Speedups (and other algorithmic advantages) resulting from the quantum world are also investigated in the context of robot planning (Chapter 4), machine learning (Chapter 5), and robot controls and perception (Chapter 6). Our survey explores how algorithms commonly used for robots are expected to change when implemented with quantum mechanisms. We survey the literature for time and space complexity differences, key changes in underlying properties, and possible tradeoffs in scaling commonly used robotic techniques in quantum media. Our book also highlights some of the current implementations of quantum engineering mechanisms (Chapter 7) as well as current limitations. Finally, we conclude with a holistic summary of potential benefits to robotics from quantum mechanisms (Chapter 8).

1.3   QUANTUM OPERATING PRINCIPLES

Quantum approaches can be difficult to understand. Their mathematics can be quite nuanced and esoteric to the uninitiated reader. Even someone who is a talented robotics engineer and master of traditional mathematically intense robotic methods may struggle! To make quantum approaches easier to comprehend, our book boils each technique we discuss down to its essential Quantum Operating Principles (QOPs).

QOPs is a presentation style we introduce to make the assumptions of quantum approaches clearer. Many of the more sophisticated algorithms are really just applications of a few fundamental quantum principles.

Whenever we discuss a quantum improvement for a robot, we do so in relation to the classical techniques used in robotics. For the quantum technique, we attempt to highlight its fundamental QOPs and the potential advantages of the quantum technique to the classical method. At the end of each chapter, we also include a table of QOPs that different quantum methods discussed in the chapter use. We hope that these explanations will make the reader's journey into quantum robotics smoother.

Quantum robotics (and quantum computing at large) are fields whose fundamentals are still in flux. They are exciting fields with daily new insights and discoveries. However, the best ways to engineer quantum systems are still being debated. Because of the rapid movement of the field, we believe that the best student of quantum robotics is one that understands the fundamental assumptions of different methods. If tomorrow a particular quantum theory were to accumulate more evidence, the algorithms and techniques based on it would be more likely to be used in the future for robots. Conversely, if a particular quantum theory is proven false, it is good to know which techniques in the literature will not pan out. Our goal with the QOPs breakdown is to help readers understand the spectra of possible truth in the quantum world, since there is not yet certainty.

In the next section, we introduce the basics of the current theory of quantum mechanics. Later sections will apply these QOP concepts to robotic search and planning, machine learning, sensing, and controls.

Relevant Background on Quantum Mechanics

In this section, we provide a concise survey of key concepts from quantum mechanics that are essential for quantum robotics. In general, our work is written for an audience familiar with typical robotic algorithms and technologies and presumes no prior knowledge of quantum mechanics.

2.1 QUBITS AND SUPERPOSITION

The fundamental unit of quantum computation is the qubit. The qubit can be thought of as the "transistor" of a quantum computer. A classical transistor controls a single binary bit that represents just a single discrete value, 0 or 1. A quantum bit, or qubit, assumes a complex combination of the two states, 0 and 1. This leads to some special properties unique to qubits. For instance, classical bits are independent of each other. Changing the value of a classical bit generally does not affect the value of other classical bits. This is not the case with qubits. As we will see, qubits can represent exponentially more data via special properties of quantum mechanics: superposition and entanglement.

As a simple illustration of the qubit, consider an electron orbiting a nucleus in an atom. The electron can be in one of two orbital states: the "ground" state or the "excited" state. Figure 2.1 shows an example depiction of this simple case. The electron functions as a qubit, and the qubit's computational data is encoded by the electron's orbital states.

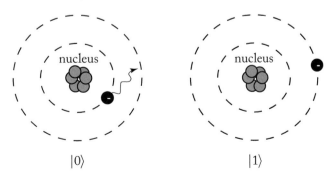

Figure 2.1: Illustration of simple qubit.

Bra-ket notation, originally invented by Paul Dirac in 1939, is a standard notation for representing states of quantum systems. A ket $|A\rangle$ represents the numeric state vector of a quantum system and is typically an element of a Hilbert space.[1] With ket notation, the ground state of our simple qubit can be represented as the ket $|0\rangle$ (an abbreviation for the state vector $\begin{bmatrix} 1 & 0 \end{bmatrix}^T$), and the excited state can be represented as the ket $|1\rangle$ (an abbreviation for the state vector $\begin{bmatrix} 0 & 1 \end{bmatrix}^T$). The bra $\langle A|$ is defined mathematically as the conjugate transpose[2] of a ket (e.g., $\langle A| = |A\rangle^{\dagger}$).

Before measurement, the electron is said to be in a superposition of the two states, denoted as a weighted sum:

$$|\psi\rangle = \alpha\,|0\rangle + \beta\,|1\rangle \tag{2.1}$$

where α and β are complex numbers. The α and β coefficients encode the probability distribution of states the electron can be found in when measured by a lab instrument. Until measurement, the true underlying state of the electron is not known. In fact, technically speaking, the true state of the electron is a linear superposition of both the ground and excited state. The superposition notation indicates that the electron is simultaneously in *both* the ground and excited state.

When the qubit is measured, its superposition collapses to exactly one state (either the ground or excited state), and the probability of measuring a particular state is given by its amplitude weights. The electron is measured to be in the ground state $|0\rangle$ with probability $|\alpha|^2$ and in the excited state $|1\rangle$ with probability $|\beta|^2$ such that $|\alpha|^2 + |\beta|^2 = 1$.

The notation can be generalized for describing k-level quantum systems. In a k-level quantum system, the electron can be in one of k orbitals as opposed to just one of two states. The state of the k-level quantum system $|\psi\rangle$ (when in superposition) can be expressed as:

$$|\psi\rangle = \sum_{i=1}^{k} \alpha_i\,|i\rangle$$
$$\text{s.t.} \quad \sum_{i=1}^{k} |\alpha_i|^2 = 1. \tag{2.2}$$

Upon measurement of the system, the superposition collapses to state $|i\rangle$ with probability $|\alpha_i|^2$. The α_i are complex numbers, potentially having both real and imaginary parts.

2.2   QUANTUM STATES AND ENTANGLEMENT

Previously, we illustrated how a simple electron-orbital system could be represented with bra-ket notation. The ket is a mathematical abstraction, a notation representing a physical state that exists

---

[1]Often, for us, just $\mathbb{C}^N$, the space of complex numbered vectors.

[2]The conjugate transpose of a matrix $A^* = \overline{A^T}$. To form the conjugate transpose of A, one takes the transpose of A and then computes the complex conjugate of each entry. The complex conjugate is simply the negation of the imaginary part (but not the real part).

in the real world. The beauty of this abstraction is that a variety of quantum systems, although implemented differently, can be described by the same underlying theory.

For a particular quantum system being studied, a physicist using the bra-ket notation will specify some of the system's elementary physical states as "pure states." Pure states are defined as fundamental states of a quantum system that cannot be created from other quantum states. A pure state $|\psi\rangle$ can be described via a density matrix:

$$\rho = |\psi\rangle \langle\psi|. \tag{2.3}$$

In general, each quantum state (pure or not) has an associated density matrix. Not all states are pure; many are mixtures of pure states. A probabilistic mixture of pure states (called a "mixed state") can be represented by the following density matrix:

$$\rho_{\text{mixed}} = \sum_s P_s |\psi_s\rangle \langle\psi_s|$$
$$\text{s.t.} \sum_s P_s = 1. \tag{2.4}$$

where $|\psi_s\rangle$ are the individual pure states participating in the mixture, and the P_s are mixing weights.

Composite systems are quantum systems that consist of two or more distinct physical particles or systems. The state of a composite system may sometimes be described as a tensor product ($\otimes$) of its components.

Here is an example of a 2-qubit system. $|\psi\rangle_A$ and $|\psi\rangle_B$ are two qubits that have probability distributions for being measured in states $|0\rangle$ and $|1\rangle$ respectively. The tensor product ($\otimes$) of their distinct probability distributions can sometimes represent the joint probability distribution of the composite system's measurement outcome probabilities.

$$|\psi\rangle_A = \begin{bmatrix} \alpha \\ \beta \end{bmatrix}$$
$$|\psi\rangle_B = \begin{bmatrix} \gamma \\ \delta \end{bmatrix}$$
$$|\psi\rangle_{AB} = |\psi\rangle_A \otimes |\psi\rangle_B \tag{2.5}$$
$$|\psi\rangle_{AB} = \begin{bmatrix} \alpha \\ \beta \end{bmatrix} \otimes \begin{bmatrix} \gamma \\ \delta \end{bmatrix} = \begin{bmatrix} \alpha\gamma \\ \alpha\delta \\ \beta\gamma \\ \beta\delta \end{bmatrix}.$$

The composite system, in this 2-qubit example, is measured in state $|00\rangle$ with probability $\alpha\gamma$, $|01\rangle$ with probability $\alpha\delta$, $|10\rangle$ with probability $\beta\gamma$, and $|11\rangle$ with probability $\beta\delta$. The tensor product operation $\otimes$ is taking in two vectors representing probability distributions of possible measurement outcomes. Each of these input probability distributions sums to 1. The output is a new vector that holds the joint probability distribution (that also sums to 1), under the critical assumption that the measurement outcome probabilities are statistically independent.

In a key feature of quantum mechanics, a composite system is often more than the sum of its parts. Groups of particles can interact in ways such that the quantum state of each particle cannot be described independently. Composite states that cannot be written as a tensor product of components are considered "entangled." For example, the composite state:

$$\frac{|00\rangle + |11\rangle}{\sqrt{2}} \overset{?}{=} (\alpha\,|0\rangle + \beta\,|1\rangle) \otimes (\gamma\,|0\rangle + \delta\,|1\rangle) \tag{2.6}$$

cannot be written as a product state of the two qubits forming the composite state. Expanding out the tensor product, we see that the system of equations $\{\alpha\gamma = \frac{1}{\sqrt{2}}, \alpha\delta = 0, \beta\gamma = 0, \beta\delta = \frac{1}{\sqrt{2}}\}$ has no solution. The entangled composite system thus cannot be decomposed into its individual parts.

When the composite system can be represented using the tensor product decomposition, the qubit measurement events are effectively statistically independent probability events. The joint measurement outcome probabilities equal the numeric multiplication of individual measurement probabilities. In Equation (2.6), the composite system consists of entangled qubits, and the statistical independence assumption no longer holds. In an entangled system, the qubits exhibit state correlations. If one knows the state of one qubit in an entangled pair, he or she necessarily obtains information about the state of the other entangled qubit.

The overall quantum state (in superposition) of a composite system with N qubits having state $|\psi\rangle$ (or more generally as N quantum subsystems each with quantum state $|\psi\rangle$) can be denoted as $|\psi\rangle^{\otimes N}$. Entanglement of qubits provides a potentially powerful data representation mechanism. Classically, N binary bits can represent only one N-bit number. N qubits, however, can probabilistically represent 2^N states in a superposition. N qubits can thus represent all possible 2^N N-bit numbers in that superposition. This advantage is partially offset by the additional processing overhead necessary to maintain quantum memory (since a qubit can only take on one state when measured and thus maintains the data only probabilistically). However, even with the additional overhead, quantum storage is expected to produce data representation advantages over classical implementations.

2.3   SCHRÖDINGER EQUATION AND QUANTUM STATE EVOLUTION

Quantum states change according to particular dynamics. The Schrödinger Equation can be used to describe a quantum system's time-evolution:

$$i\hbar\frac{\delta}{\delta t}\,|\psi\rangle = H\,|\psi\rangle \tag{2.7}$$

where $|\psi\rangle$ is the state of the quantum system, H is a Hamiltonian operator representing the total energy of the system, $\hbar$ is Planck's constant, and $i = \sqrt{-1}$. Schrödinger's Equation expresses that the time-evolution of a quantum system can be expressed in terms of Hamiltonian operators. This description of quantum systems is key to Adiabatic Optimization (see Section 2.5.4).

Hamiltonian operators that govern the evolution of quantum systems have special structure. In general, the evolution of a closed quantum system must be unitary, and the time-evolution of a closed quantum system can be described by application of a sequence of matrices that are unitary.

Formally, a complex square matrix U is unitary if its conjugate transpose $U^\dagger$ is also its inverse. This means:

$$U^\dagger U = UU^\dagger = I \tag{2.8}$$

where I is the identity matrix. Unitary matrices have several useful properties including being norm-preserving (i.e., $\langle Ux, Uy \rangle = \langle x, y \rangle$ for two complex vectors x and y), being diagonalizable (i.e., writable as $U = VDV^*$), having $|\det U| = 1$, and being invertible.

Unitary matrices help formalize the evolution of a quantum system. The state vector $|\psi\rangle$ of a quantum system can be pre-multiplied by a unitary matrix U. When a state vector containing a probability distribution over measurement outcomes is pre-multiplied by a unitary matrix, the operation always produces a new probability distribution vector whose elements also sum to 1. The resulting probability distribution represents the possible measurement outcomes of the quantum system after the system is evolved by the unitary matrix operator U. Unitary matrix operators can also be chain multiplied together (e.g., $U_1 U_2 \ldots U_n$) to represent a sequence of evolution steps on a quantum system. As we will see next, another view of unitary matrices is as logic gates in a quantum circuit that process input data (i.e., quantum states) and return outputs.

2.4 QUANTUM LOGIC GATES AND CIRCUITS

Quantum logic gates are the analogue to classical computational logic gates. Computational gates can be viewed as mathematical operators that transform an initial data state to a final data state. Since quantum state evolution must be unitary, quantum gates must be unitary operators.

2.4.1 REVERSIBLE COMPUTING AND LANDAUER'S PRINCIPLE

An interesting departure from classical computation is that quantum computer gates are always reversible. One can always, given the output and the operators, recover the initial state before the computation. This follows because a unitary matrix used to evolve a quantum system is also invertible.

Logic reversibility is the ability to determine the logic inputs by the gate outputs. For example, the classical NOT gate is reversible, but the classical OR gate is not. By definition, a reversible logic circuit has the additional following properties [Vos, 2010]:

1. The number of inputs and outputs are the same in the circuit.

2. For any pair of input signal assignments, there are two distinct pairs of output signal assignments.

Conveniently, the truth table of a reversible circuit with width n is represented as a square matrix of size 2^n. While there are $(2^n)!$ different Boolean logic circuits of width n that can be realized, only a handful are valid reversible computing mechanisms.

Reversible quantum computing has a remarkable thermodynamic interpretation. In a physical sense, a reversible circuit will preserve information entropy, i.e., lead to no information content lost. Landauer's principle [Landauer, 1961] states that any logically irreversible manipulation of information (such as the erasure of a classical digital bit) must lead to entropy increase in the system. The principle suggests the "Landauer Limit" that the minimum possible amount of energy required to erase one bit of information is $kT \ln (2)$ where k is Boltzmann's constant and T is the temperature of the circuit. At room temperature, the Landauer Limit suggests that erasing a bit requires a mere 2.80 zettajoules!

The energy expenditure of current computers is nowhere near this theoretical limit. The most energy-efficient machines today still use millions of times this forecasted energy amount. In fact, in many realms of computer science and engineering, there is an expectation of intelligent computation being a highly power-intensive activity. Even neuroscientific predictions estimate human brain activities account for more than 20% of the body energy needs, with more than two-thirds of power consumption associated with problem solving activities [Swaminathan, 2008]. It seems natural to expect intelligence to be power-intensive. At the same time, the brain only uses about 20W of electricity, which is less than the energy required to run a dim light bulb [Swaminathan, 2008]. Clearly, more can still be done to optimize power consumption of digital circuits, even if generating intelligent behavior requires more energy consumption than other useful functions.

Excitingly, many studies appear to confirm the Landauer predictions for small-scale circuitry (though, convincing empirical proof is not without counterargument). Bennett [1973] showed the theoretical validity of implementing an energy efficient reversible digital circuit in terms of a three-tape Turing machine. In 2012, an experimental measurement of Landauer's bound for the generic model of one-bit memory was demonstrated empirically [Bérut et al., 2012]. Recently, Hong et al. [2016] used high precision magnetometry to measure the energy loss of flipping the value of a single nanomagnetic bit and found the result to be within tolerance of the Landauer limit (about 3 zettajoules).

Since bulky battery technology is one of the key limiting factors of many current robotic systems, Landauer's Principle provides hope for increasing the computational power of robots while simultaneously making robots more power-efficient. If true, Landauer's Principle suggests a world with highly energy-efficient robots operating with quantum-scale circuits that allow massive reduction in power consumed.

2.4.2   NOTABLE QUANTUM GATES

In this section, we describe some of the common quantum gates used in quantum circuits. Since all operations that change a quantum system are unitary, all quantum gates must, by definition, be unitary matrices.

The logical NOT gate can be described by the following unitary matrix:

$$N = \begin{bmatrix} 0 & 1 \\ 1 & 0 \end{bmatrix}. \tag{2.9}$$

This matrix converts the quantum state $\alpha |0\rangle + \beta |1\rangle$ to $\alpha |1\rangle + \beta |0\rangle$ (via matrix multiplication). A gate useful throughout quantum computation is the Hadamard gate:

$$H = \frac{1}{\sqrt{2}} \begin{bmatrix} 1 & 1 \\ 1 & -1 \end{bmatrix} \tag{2.10}$$

which converts $|0\rangle \rightarrow \frac{|0\rangle + |1\rangle}{\sqrt{2}}$ and $|1\rangle \rightarrow \frac{|0\rangle - |1\rangle}{\sqrt{2}}$. Because of its effect, the Hadamard gate is sometimes called the "half way" gate. The Hadamard gate is an important building block for quantum parallelism (discussed in Section 2.5.1).

The Controlled Not (CNOT) gate converts $|X_1, X_2\rangle \rightarrow |X_1, XOR(X_1, X_2)\rangle$. Effectively, this gate negates X_2 if X_1 is set to 1. The graphic representation of CNOT and its unitary matrix are shown in Figure 2.2.

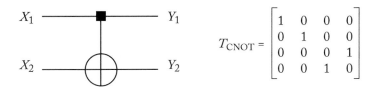

Figure 2.2: Controlled NOT (CNOT) gate—Depending on the value of X_1, the value of X_2 may be flipped.

The "swap operation" (used in algorithms such as the quantum dot product described in Section 5.1.2 is created from three CNOT gates and converts $|X_1, X_2\rangle \rightarrow |X_2, X_1\rangle$.

For a three qubit network, the well-known Tofolli gate converts $|X_1, X_2, X_3\rangle \rightarrow |X_1, X_2, XOR(X_3, X_1 \text{ AND } X_2)\rangle$, flipping the third input X_3 if both X_1 and X_2 are set to 1.

Similarly, the Fredkin gate performs a controlled swap on three qubits, i.e., if $X_1 = 1$, the output state $Y_{1:3}$ is $Y_1 = X_1$, $Y_2 = X_3$, and $Y_3 = X_2$. Otherwise, $Y_1 = X_1$, $Y_2 = X_2$, and $Y_3 = X_3$.

The basic gates described here are useful for building more advanced quantum circuits.

2.4.3   QUANTUM CIRCUIT FOR FAST FOURIER TRANSFORM

The Fast Fourier Transform (FFT) plays an important role in classical image processing, pattern recognition, and signal processing. Hence, the FFT is likely to see implementation in quantum circuits for robots.

The classical FFT is a fast way to compute the classical Discrete Fourier Transform (DFT). The DFT calculates N coefficients of complex sine waves to model N complex samples of an unknown function. If $A = \{a_1, \dots, a_N\}$ is the input data and $P = \{p_1, \dots, p_N\}$ is the output data, DFT solves the linear equation $P = F \times A$, where F is an $N \times N$ constant matrix with complex numbered elements $F_{x,y} = \exp(-2\pi i / N)^{(x-1)(y-1)}$. The solution requires $O(N^2)$ operations. Classical FFT exploits unique properties of the F matrix to reduce the calculations to $O(N \log_2(N))$ operations.

A key difference between the Quantum Fast Fourier Transform (QFFT) and the classical FFT is that QFFT can be implemented in fewer circuit elements. For N input qubits, the QFFT equation is $P' = F' \times A'$, where $A' = \{a_1, \dots, a_{2N}\}$, $P' = \{p_1, \dots, p_{2N}\}$ and $F'_{x,y} = \exp(-2\pi i / 2^N)^{(x-1)(y-1)}$.

Conveniently, the realization of this form requires only $N(N + 1)/2$ quantum gates [Vos, 2010]. Hadamard and Controlled phase gates can be used to construct QFFT circuits. The controlled phase gate implements a phase shift of the second qubit (as shown in Figure 2.3). In Figure 2.4, we see a quantum circuit realization for the Quantum Fast Fourier Transform for $N = 3$.

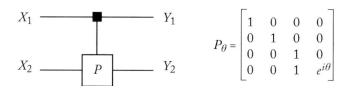

$$P_\theta = \begin{bmatrix} 1 & 0 & 0 & 0 \\ 0 & 1 & 0 & 0 \\ 0 & 0 & 1 & 0 \\ 0 & 0 & 1 & e^{i\theta} \end{bmatrix}$$

Figure 2.3: Controlled phase gate. Depending on X_1, the value of X_2 is flipped.

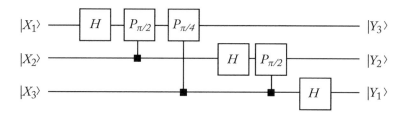

Figure 2.4: Quantum FFT, $N = 3$.

2.5 QUANTUM COMPUTING MECHANISMS

In general, quantum gates placed in a circuit can create interesting computational behavior not possible in classical circuits. Quantum computing methods offer potential speedups and improved properties for many classical algorithms. Some introductory background can be found in Kaye et al. [2006], Rieffel and Polak [2000], Steane [1998]. The key general mechanisms of quantum speedup and improvement are discussed in this section.

2.5.1 QUANTUM PARALLELISM

The first key mechanism for algorithmic speedup is quantum parallelism. A parallelized implementation of a function computes its value on multiple inputs *simultaneously*. By using quantum gates, one can create circuits that exhibit parallelism capability not possible in classical circuits. One can see the basic pattern in the following quantum circuit (shown in Figure 2.5).

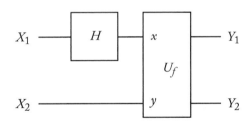

Figure 2.5: Illustration of quantum parallelism circuit.

Let the goal be to parallelize the function $f(x) : \{0, 1\} \to \{0, 1\}$. We define the unitary circuit gate U_f for that function as follows.

$$U_f \, |x, y\rangle := |x, y \oplus f(x)\rangle. \tag{2.11}$$

The x circuit line is known as the data register, and the y circuit line is known as the target register. Figure 2.5 puts together the designed U_f gate and a Hadamard gate.

Fix the inputs to circuit at $X_1 = |0\rangle$ and $X_2 = |0\rangle$. The Hadamard gate produces the output $(\frac{|0\rangle+|1\rangle}{\sqrt{2}})$, which is then fed into the data register (x) of the U_f gate. The U_f gate evaluates its function at $x = (\frac{|0\rangle+|1\rangle}{\sqrt{2}})$, the output of which forms a composite system with $y = |0\rangle$. The circuit produces the following pattern:

$$U_f\left(\left(\frac{1}{\sqrt{2}}\,|0\rangle + \frac{1}{\sqrt{2}}\,|1\rangle\right)\otimes|0\rangle\right) = \frac{1}{\sqrt{2}}U_f\,|0\rangle\,|0\rangle + \frac{1}{\sqrt{2}}U_f\,|1\rangle\,|0\rangle$$

$$= \frac{1}{\sqrt{2}}U_f\,|0,0\rangle + \frac{1}{\sqrt{2}}U_f\,|1,0\rangle$$

$$= \frac{1}{\sqrt{2}}\,|0,0\oplus f(0)\rangle + \frac{1}{\sqrt{2}}\,|1,0\oplus f(1)\rangle \tag{2.12}$$

$$= \frac{1}{\sqrt{2}}\,|0,f(0)\rangle + \frac{1}{\sqrt{2}}\,|1,f(1)\rangle$$

$$= \frac{|0,f(0)\rangle + |1,f(1)\rangle}{\sqrt{2}}.$$

Thus, the circuit evaluates the function $f(x)$ at both 0 and 1 simultaneously in parallel. It turns out that this pattern generalizes to n qubits for a U_f gate that has n data registers. Applying a Hadamard gate to each data register creates the famous pattern known as the "Hadamard transform." For an input state, $|0\rangle^{\otimes n}\,|0\rangle$, this pattern will generate the state:

$$|\psi\rangle = \frac{1}{\sqrt{2^n}}\sum_{x\in X}|x\rangle\,|f(x)\rangle \tag{2.13}$$

where X is the set of 2^n unique binary bit strings of length n. The Hadamard Transform produces a superposition of 2^n states using n gates, simultaneously evaluating all values of the function while the system is in superposition. This is a remarkable amount of computation done, compared to classical systems.

2.5.2 CHALLENGES WITH QUANTUM PARALLELISM

The major limitation of quantum parallelism is that, while the output state contains information about multiple function evaluations of $f(x)$ when the system is in superposition, only one of the function evaluations is returned when the quantum system is measured. In the basic Hardamard transform, any of the function values $|f(x)\rangle$ can be returned, each with equal probability. Limited data return from measurement is a common theme in quantum computing. For many methods, it is common for a quantum system to perform a significant amount of computation while in superposition but to reveal only a small fraction of the result upon measurement.

Much research is dedicated to evolving quantum systems that encode useful information in superposition so that the most useful output information is returned upon measurement. For example, there exist quantum parallel techniques to extract some meaningful information about the global properties of a function as opposed to just raw function values at points. Deutsch's algorithm [Deutsch and Jozsa, 1992] allows for the evaluation of the parity $f(0)\oplus f(1)$ using one parallel computation whereas classically two computations would be required.

In general, a convincing empirical demonstration of quantum parallelism is a subject of ongoing research. Some authors ask whether quantum parallelism (in the unitary circuit formulation) is even realizable. The construction of the U_f gate in practice may require internal logic that scales with the number of unit-cost operations for the precision of x. The quantum circuit complexity result may be benefiting from gate-level parallelism not realizable in classical circuits. When one looks at the complexity of the complete circuit, there may not always be an advantage of quantum parallelism to classical parallelism [Lanzagorta and Uhlmann, 2008b].

Despite the current challenges with quantum parallelism, it is still regarded as a major hope for the future success of quantum computing. It is also one of the fundamental pillars on which more sophisticated quantum algorithms are built.

2.5.3   GROVER'S SEARCH ALGORITHM

Grover's search algorithm, which makes use of quantum parallelism, is a method for quantum computers to find an element in an unordered set of N elements quadratically faster than the theoretical limit for classical algorithms. The algorithm is expected to run in $O(\sqrt{N})$ time as opposed to $O(N)$ time classically.

In order to do this, the algorithm uses a quantum oracle, a black box able to indicate the solution for a given computational problem. While this sounds somewhat magical, the concept is quite logical. A quantum oracle is a function $f(x)$ that is able to look at a possible solution to a computational problem and verify that it is a solution. For many computational problems, it is intensive to compute a solution but easy to verify one. The quantum oracle does not have to compute the solution, just to know when one is found.

A quantum oracle function $f(x)$ is defined for the data that indicates if the sought element is found:

$$|x\rangle |q\rangle \xrightarrow{O} |x\rangle |q \oplus f(x)\rangle \tag{2.14}$$

q is an ancilla bit that is flipped if $f(x) = 1$ and stays the same if $f(x) = 0$.

How many calls to the oracle are needed to find the sought element? Classically, $O(N)$ calls to the abstract oracle are required since, in the worst case, the entire database may need to be searched. However, with a quantum computer, only $O(\sqrt{N})$ calls to the oracle are needed to find the sought element.

If there are N items, an item's binary index can be represented with $n = \log_2(N)$ bits. A Hadamard transform is used on $|0\rangle^{\otimes n}$ to prepare the equally-weighted superposition over the N items in the database:

$$|\psi\rangle = \frac{1}{\sqrt{2^n}} \sum_{x=0}^{2^n-1} |x\rangle. \tag{2.15}$$

The Grover operator is defined as:

$$G = (2 |\psi\rangle \langle \psi| - I) O \tag{2.16}$$

where O is the oracle function and I is the identity matrix. After transforming the system with the Grover operator $O(\sqrt{N})$ times, measurement of the resulting system will, with high probability, correspond to the index of the found item. The complete Grover's algorithm is given below.

Algorithm 2.1 Grover's Search Algorithm

Input : Initial State $|0\rangle^{\otimes n}$, Requested element x^*
Output : Database index of x^* (if x^* exists in database)

1: Initialize equally-weighted superposition $|\psi\rangle$.
2: **for** $O(\sqrt{N})$ times **do**
3:    Apply the Grover operator G to $|\psi\rangle$.
4: **end for**
5: Measure the system.
6: **return** With high probability, the index of x^* will be the measurement result.

The intuition behind Grover's algorithm is shown in Figure 2.6. All elements originally start in a superposition where each element has the equal weight. Each application of the Grover operator (which contains the oracle function) inverts the phase of solution elements x^*, while keeping the elements of non-solution elements the same. The Grover operator then inverts solution elements x^* above the mean.

As the Grover operator is applied multiple times, the amplitude weights of the solution elements become larger and larger. When the Grover operator is applied a sufficient number of times, the solution x^* is obtained with extremely high probability when the quantum system is measured.

For an N-item database with 1 solution item, Grover's algorithm will find the solution in $O(\sqrt{N})$ time. If there are t (identical) solution items, the time complexity is $O(\sqrt{\frac{N}{t}})$ [Boyer et al., 2004].

The fine print of Grover's algorithm complexity is that these results only take into account the number of calls to the oracle function, not the complexity of implementing the oracle function circuit. Note that for many useful functions (such as those used in encryption), there exist relatively straightforward verification oracles. The oracle can check whether a solution exists, without going into the details of how the solution is computed. While the solution can be difficult to compute, if the oracle can verify the solution, it is often a sufficient condition for Grover's algorithm to solve the search problem effectively.

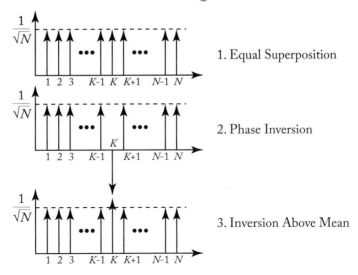

Figure 2.6: Intuition behind mathematical principles of Grover's algorithm.

2.5.4   ADIABATIC QUANTUM OPTIMIZATION

Many search problems can also be viewed as optimization problems. For example, many search problems can be cast with the objective of finding the global minimum (x_0, f_0) of a given function $f(x)$:

$$
\begin{aligned}
\min_x f(x) &= f_0 \\
\operatorname*{argmin}_x f(x) &= x_0.
\end{aligned}
\tag{2.17}
$$

Adiabatic Quantum Optimization, a particular approach to quantum computing, uses this formulation as its main mathematical machinery. An adiabatic process is one that changes conditions gradually so the system can adapt to its configuration. The adiabatic theorem [Born and Fock, 1928] describes the time-evolution of the Hamiltonian of an adiabatic process:

$$
H(\lambda(t)) = (1 - \lambda(t))H_0 + \lambda(t)H_1.
\tag{2.18}
$$

The Hamiltonian H_0 is a Hamiltonian whose ground state can easily be constructed in the lab. The Hamiltonian H_1 is one whose ground state energy is the target. The adiabatic theorem states that if the system starts in the ground state of $H(0) = H_0$ and the parameter $\lambda(t)$ is slowly increased, the system will evolve to the ground state of H_1 if $H(\lambda(t))$ is always gapped (i.e., there is no degeneracy for the ground state energy).

The adiabatic theorem forms the basis for adiabatic quantum computing and the design of adiabatic processors, though the question of how much speedup is achieved is complex [Rønnow et al., 2014]. The annealing time in the adiabatic theorem depends on the eigenvalue gap, which is not known *a priori*. Depending on how the eigenvalue gap scales with the problem size, the annealing time could grow anywhere from polynomially to exponentially. The speedup also depends on the classical algorithms that serve as benchmarks. Different classical algorithms may perform better or worse on different problems. Thus, it is sometimes difficult to decide what is a fair comparison. For many problem instances, quantum annealing is forecasted to offer roughly a quadratic factor speedup compared to many classical stochastic optimization approaches.

2.5.5   ADIABATIC HARDWARE AND SPEEDUPS

Quantum adiabatic optimization leverages stochastic quantum annealing procedures in hardware to produce speedups in optimizing robotic algorithms. In this section, we describe how algorithmic problems are mathematically mapped onto an adiabatic processor. Various implementations of adiabatic hardware exist and are discussed further in Chapter 7.

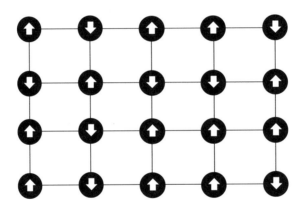

Figure 2.7: Illustration of Ising model.

The Ising model from statistical mechanics studies the magnetic dipole moments of atomic spin. Imagine a lattice of elements that can spin up or down (as shown in Figure 2.7). In the model, only neighboring pairs of elements are assumed to interact. The elements choose spins to minimize the lattice's overall energy configuration:

$$\underset{s}{\text{argmin}} \left(s^T J s + h^T s \right) \tag{2.19}$$

where $s_i \in \{-1, +1\}$ are variables representing "spins," J is a matrix describing interactions betweens spins, and h is a numeric vector that models the impact of an external magnetic field. The Ising objective function is represented by the following Hamiltonian:

$$H_I = \sum_{i,j} J_{i,j} \sigma_i^z \sigma_j^z + \sum_i h_i \sigma_i^z \qquad (2.20)$$

where i and j are pairs of adjacent sites and σ_i^z is the spin variable of qubit i.

Define the start state of the adiabatic process as:

$$H_B = \sum_i \left(\frac{1 - \sigma_i^x}{2} \right). \qquad (2.21)$$

The adiabatic theorem (discussed in Section 2.5.4) can be used to solve the energy minimization problem:

$$H(\lambda(t)) = (1 - \lambda(t))H_B + \lambda(t)H_I. \qquad (2.22)$$

Quadratic Unconstrained Binary Optimization (QUBO) is the mathematical optimization form of a simplified Ising model:

$$\min x^T Q x \quad \text{s.t.} \quad x \in \{0, 1\}^n \qquad (2.23)$$

where x is an $n \times 1$ vector of Boolean variables, and Q is an $n \times n$ square symmetric matrix of coefficients.

If an optimization problem can be cast as a QUBO, it can be represented as an Ising model and optimized by adiabatic hardware. Many computational problems can be cast as QUBOs. Abstractly, the Ising model represents a set of elements and their pair-wise interactions. The Ising model has seen many real-world modeling applications because interacting elements must be modeled in a large number of fields. The world is full of interacting elements (e.g., electrons in an atom, neurons in a network, birds in a flock, or even interacting social agents in a crowd psychology setting).

Current adiabatic hardware is mostly limited to solving QUBOs. As such, it is, at best, a limited form of universal quantum computing. In addition, current hardware does not allow many qubits, produces many defects such that not all pairs of qubits are entangled, and must use sparse, pre-specified topologies for qubit interconnections [Neven et al., 2009, Pudenz et al., 2014].

2.5.6 SHOR'S QUANTUM FACTORIZATION ALGORITHM

One of the hallmark results in quantum computing is Shor's algorithm for integer factorization. When first proposed, Shor's algorithm was one of the key theoretical insights that showed quantum computing could be very different from (and has the potential to be significantly more powerful than) classical computing.

Integer factorization is a fundamental problem in computer science. While it is relatively easy to find large prime numbers, it is particularly hard to resolve a composite number into its prime factors. A decomposition of an arbitrary N-bit number, executed on a classical computer, has an exponential complexity $O(k^N)$, for some constant k. Prime factorization is among the class of NP problems.

The hardness of factorization has historically served as a boon in fields such as public-key cryptography and network communication. The widely used RSA encryption scheme [Rivest et al., 1978] has been one of the more popular algorithms for secure communication on networks [Menezes et al., 1996]. In a robotics context, algorithms such as RSA help keep robot communication and teleoperation secure from possible hackers. Quantum factorization, however, could be used to break RSA and change the nature of internet security.

In 1994, Peter Shor formulated a hallmark quantum factorization algorithm that performs factorization in polynomial time [Shor, 1999]. For an arbitrary odd number N, consider a random x, $1 < x < N$, for which $x^r = 1 \mod N$, for some r. The series $(1, x, x^2, x^3, \dots) \mod N$ is periodic with a period not greater than N. The above relations can be succinctly represented as:

$$(x^{r/2} - 1)(x^{r/2} + 1) = 0 \mod N. \tag{2.24}$$

From the equation $(ab = 0 \mod N)$, one can find, in polynomial time, the greatest common divisors, $gcd(a, N)$ and $gcd(b, N)$. These divisors, if they are found, will be factors of N. If a non-trivial divisor does not exist, the x variable can be re-picked and the computation repeated.

The described steps can be performed on a classical computer in polynomial time, with the exception of calculating the exponential function x^r. Shor's algorithm uses the Discrete Fourier Transform (DFT) and quantum parallelism to calculate the periodic function simultaneously for many values x, so that all operations happen in $O(N)$ time.

Shor's algorithm is a major advancement in the theory of quantum computing, and one of the most interesting predictions posited by the field. A practical implementation for large integer factorization instances is still a subject of major research [Dattani and Bryans, 2014].

2.5.7 QUANTUM TELEPORTATION

Another major engineering possibility posited by quantum mechanics is quantum teleportation. Quantum teleportation is a technique which transmits a quantum state by using a pair of entangled qubits and a classical communication channel. In this section, we provide a brief outline of how quantum teleportation works.

Consider two parties, Alice and Bob, who wish to communicate the state $|\psi\rangle$ of a qubit. Alice wants to send the qubit state to Bob. However, Alice does not know what the state of the qubit is and, by the laws of quantum mechanics, observing the state collapses the qubit into a definite value, $|0\rangle$ or $|1\rangle$. Furthermore, even if Alice knows the superposition state $|\psi\rangle = \alpha |0\rangle + \beta |1\rangle$, communicating it over a quantum channel would take an infinite amount of time, since the amplitudes α and β are complex numbers with infinite precision. However, the quantum teleportation technique allows one to transmit the state $|\psi\rangle$ using an entangled pair of bits and communication of two bits of information over a classical communication channel.

The protocol for transmitting the state $|\psi\rangle$ begins with forming an entangled pair of qubits A and B. The qubit A is given to Alice and the qubit B is given to Bob. Alice now takes the qubit $|\psi\rangle$ that she wishes to communicate and makes a joint measurement of it with the entangled qubit A. This results in Alice entangling her two qubits and observing one of four states. This observed state can be encoded in two classical information bits. Alice encodes her observations into two classical bits and transmits them to Bob. As a result of Alice's observation, B is now in one of four possible states, which correspond to the state $|\psi\rangle$. Using the information Alice sent with two classical bits, Bob can perform operations on B so that its state becomes $|\psi\rangle$. Bob can thus successfully recreate the state $|\psi\rangle$ on his end.

Quantum teleportation has many potential applications in robotics. One could potentially use quantum teleportation to transmit information about qubits between a robot and a server operating in different locations. Note that information cannot be transmitted faster than the speed of light, since the communication of two classical bits of information is needed in the protocol. One does not expect a better network speed from quantum teleportation. However, quantum teleportation might one day allow qubit information to be transported with less effort than is currently required. Some additional applications in robotics include building noise-resistant quantum gates, quantum error correcting codes, and further development of quantum information theory.

2.6   QUANTUM OPERATING PRINCIPLES (QOPS) SUMMARY

We have discussed some key basic principles of quantum mechanics and quantum computation. Many techniques in quantum robotics can be understood as applications of these core principles which we call "Quantum Operating Principles" (or QOPs). In Table 2.1, key QOPs are mentioned as well as some of their potential applications in quantum robotics that will be discussed in upcoming chapters.

2.7   CHAPTER SUMMARY

The material on fundamental quantum mechanics forms the backbone of our discussion of quantum robotics. Entanglement of qubits is the basis of quantum parallelism, a key speedup strategy for algorithms. Quantum adiabatic processors allow one to cast certain classical optimization problems in the framework of quantum optimization. In the next section, we will apply these concepts to robotic search and planning.

Chapter Key Points

- The fundamental unit of quantum computation is the qubit.

- When placed in superposition, a qubit can represent a probability distribution of values.

- When entangled with other qubits, the composite system can represent exponentially more values than classical bits.

- The Schrödinger Equation describes the unitary time-evolution of quantum systems.

- Quantum gates must be representable by unitary matrices and thus are always reversible.

- Landauer's principle suggests a bound on the minimum amount of energy required to erase one bit of information.

- The quantum Fast Fourier Transform may have a simpler circuit than the classical Fourier Transform.

- Quantum parallelism may provide exponentially more parallelism than classical parallelism (if true).

- Grover's search algorithm may allow quantum computers to find an element in an unordered set of N elements in $O(\sqrt{N})$ time, compared with $O(N)$ time required for the worst-case classical solution.

- Adiabatic optimization, based on Ising models, leverages stochastic quantum annealing procedures to produce possible speedups for certain optimization problems.

- Shor's factorization algorithm suggests possible faster factorization of a number using quantum methods compared to classical approaches.

- Quantum teleportation may allow easier transport of qubit information.

Table 2.1: Summary of Quantum Operating Principles (QOPs) and application areas in Quantum Robotics

Quantum Operating Principle	Potential Quantum Robotic Applications
Quantum Measurement	Quantum Machine Learning (Ch. 5), Quantum Hidden Markov Model (Ch. 6), Quantum Kalman Filter (Ch. 6)
Quantum State Evolution	Quantum Markov Decision Process (Ch. 4), Quantum Partially Observable Markov Decision Process (Ch. 4), Quantum Filtering and Control (Ch. 6)
Grover's Algorithm	Grover Tree Search (Ch. 3), Quantum Machine Learning (Ch. 5)
Adiabatic Theorem	Quantum STRIPS (Ch. 3), Quantum Machine Learning (Ch. 5)

C H A P T E R   3

Quantum Search

Computational search and optimization problems are ubiquitous in robotics. Computer vision has the classical challenge of scene segmentation, a search through possible visual interpretations of the world. Robot planning has the fundamental challenge of searching through sequences of plan operators to formulate plans for robot action in an environment. Classically, many planning problems are NP-complete or even PSPACE, which makes general purpose planning difficult. Robot (Machine) Learning often requires searching through feature spaces extracted from data, optimizing parameters of a model, and processing large amounts of data. Controls and manipulation have their own share of search and optimization challenges: searching through possible robot configuration spaces, optimizing kinematic settings for manipulators, and reasoning over environment dynamics.

A fundamental improvement in search algorithms can likely improve intelligence capabilities in a large number of traditional robotic challenges. Many classical robotic algorithms are expected to receive quadratic or exponential speedups from advances in quantum search.

In the quantum literature, Grover's algorithm and Adiabatic Quantum Optimization are principal mechanisms of speedup for search algorithms. The vanilla version of Grover's algorithm (previously covered in Section 2.5.3) involves search for elements in an unsorted list and requires quadratically fewer queries in the worst case than classical exhaustive search. Attempts have been made to extend Grover's algorithm to search over trees and graphs both with and without heuristic information [Tarrataca and Wichert, 2011]. The case where heuristic information is not available is known as "uninformed" search, while the use of heuristic information makes searches "informed." This chapter will highlight several of these approaches.

Additionally, quantum annealing techniques have been applied to various difficult search, navigation, and planning problems. The methods have been evaluated on current quantum adiabatic hardware [Rieffel et al., 2015, Smelyanskiy et al., 2012].

3.1   UNINFORMED GROVER TREE SEARCH

Grover Tree Search [Tarrataca and Wichert, 2011] extends Grover's algorithm to search over paths in a tree structure. We first consider the "uniformed" search case, where heuristic information is not available for the tree search.

Imagine one has a tree structure like the one shown in Figure 3.1. In a robot planning context, the nodes in the tree could correspond to possible robot-environment states and the links between nodes correspond to decisions or actions the robot can execute at each state node.

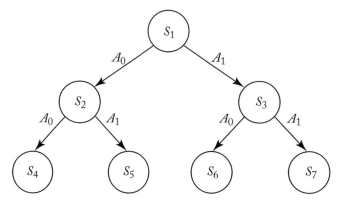

Figure 3.1: Example search tree over sequential actions for robotic planning.

Taking an action A_i at a parent state node S_j takes the robot to a particular child state node S_k in the tree. The aim of a robot planning problem may be to find a path from the start (root) node of the tree to a particular goal state node somewhere else in the tree.

A path from the root node to a child node in the tree can be represented using a bit string specifying the robot's action sequence. For example, to get from state S_1 to state S_4, the robot must choose action A_0 at S_1 to get to state S_2 and then action A_0 again at state S_2 to transition to state S_4. The bit string "00" can be used to indicate this sequence of actions taken. In similar fashion, the path from state S_1 to state S_6 can be represented by the bit string "10."

For a path of depth d, the action identifier string is d elements long. For a tree with constant branching factor b, the tree will have b^d paths to depth d. If each element requires n bits to represent, every action identifier string requires nd bits.

Consider the equal weight superposition:

$$|\psi\rangle = \frac{1}{\sqrt{b^{nd}}} \sum_{x=0}^{b^{nd}-1} |x\rangle. \tag{3.1}$$

The items in the database are a listing of all possible action identifier strings, each representing a possible path through the tree. The sought element is the action string that specifies a path from the start to a goal node. Grover's algorithm can be applied to this equal weight superposition to find the element in the database corresponding to the solution path.

Applying Grover's algorithm for $O(\sqrt{b^d})$ iterations returns the solution path with high probability when the quantum system is subsequently measured. Classical tree search, in the worst case, requires a brute force search over all $O(b^d)$ paths. Grover Tree Search thus provides a quadratic speedup (in the worst case) over classical tree search when the branching factor is constant.

Non-constant branching factor can pose a challenge for Grover Tree Search. Grover's algorithm operates on an exponentially generated database of all possible action bit strings. When the branching factor is not constant, the full superposition database will contain paths that do not actually exist in the tree. Pruning these infeasible paths requires additional book-keeping. In fact, when there are many infeasible paths, Grover Tree Search does not always provide a speedup over a brute force classical tree search. For further details, we recommend reading [Tarrataca and Wichert, 2011], which provides a rigorous complexity analysis for when Grover Trees Search speedups hold in comparison to classical uninformed tree search.

3.2   INFORMED QUANTUM TREE SEARCH

Heuristic information, when available, can provide practical speedups for many search problems on classical computers. In classical Artificial Intelligence, cost functions of the following form are often considered for informed search [Russell and Norvig, 2003]:

$$f(n) = g(n) + h(n). \tag{3.2}$$

Here $g(n)$ represents the search cost incurred to reach a particular node n, and $h(n)$ is a heuristic estimate to reach the goal from node n. This type of cost function forms the principle behind classical A^* and other such informed-search algorithms. If $h(n)$ is an admissible heuristic (that is, if it never overestimates the true cost to the goal), then such algorithms will be optimal.

Using Grover's algorithm, search problem candidate solutions are represented with qubits, and the probability of a particular candidate solution being returned is encoded in the weights of the superposition of qubits. One can incorporate some heuristic information by defining operators that act on quantum states to modify the weights in the superposition [Tarrataca and Wichert, 2011].

A key challenge of incorporating heuristic information into Grover Tree Search (or other search algorithms) is that state evolution in quantum mechanics has to be unitary. Recall that a unitary matrix U must satisfy $U^\dagger U = U U^\dagger = I$. To implement the state evaluation function $f(x)$ in the quantum world, it must be represented as a unitary operator.

Since the heuristic function $h(x)$ estimates remaining distance to the goal, one can consider states whose $f(n) < T$ (a threshold) and remove others from the search. Using this idea, one can define the unitary operator:

$$U |b\rangle |a_1 a_2 \ldots a_d\rangle = \begin{cases} - |b\rangle |a_1 a_2 \ldots a_d\rangle & \text{if } f(b, a_1, a_2, \ldots, a_d) \leq T \\ |b\rangle |a_1 a_2 \ldots a_d\rangle & \text{otherwise.} \end{cases} \tag{3.3}$$

This unitary operator considers an input action bit string that is represented as qubits, $|a_1 a_2 \ldots a_d\rangle$ up to a particular depth d. The operator sets the value of a flag qubit $|b\rangle$. The phase of $|b\rangle$ is maintained if the cost function value is above a threshold, and the phase is flipped otherwise. Heuristic information enters the search process through the setting of the threshold T for different nodes and paths. The choice of T is an important parameter for U, and how to choose it in the general case remains to be explored. In practice, Tarrataca and Wichert [2011] suggest using quantiles or other statistical approaches on known properties of the search space.

The proposed method of encoding heuristic information in quantum search has different optimality properties from classical A^*, depending on choice of T. In addition, it does not circumvent the problem of comparing multiple paths. Using this approach for encoding heuristic information into Grover Tree Search, some speedups have been proven analytically for some 8-puzzle problem instances [Tarrataca and Wichert, 2011].

3.3   APPLICATION OF QUANTUM ANNEALING TO STRIPS CLASSICAL PLANNING

Another approach to quantum search is utilization of quantum adiabatic hardware. Quantum annealing approaches have been developed to help address difficult search and classical planning problem frameworks such as STRIPS [Rieffel et al., 2015, Smelyanskiy et al., 2012].

3.3.1   CLASSICAL STRIPS PLANNING

STRIPS [Fikes and Nilsson, 1971] is a classical framework (developed in 1971 at SRI International) for robot planning problems. STRIPS is grounded in propositional logic that allows a robot to reason logically about what actions it can take from a start state to reach an end goal state.

A STRIPS instance consists of an initial specified state for the robot/environment, a set of goal states the planner is trying to reach, and a set of actions a robot may take. The planner plans from the initial state to one or many goal states using the set of actions. Each action consists of a set of binary preconditions specifying the state of the world/robot that must be true before the action can be taken and a set of binary postconditions specifying what is true for the world/robot after the action is taken.

Frameworks such as Planning Domain Definition Language (PDDL) allow specification and solution of STRIPS-encoded planning problems of various sizes [McDermott et al., 1998]. In general, however, deciding whether any plan exists for a STRIPS problem instance is PSPACE-complete [Bylander, 1994].

3.3.2   APPLICATION OF QUANTUM ANNEALING TO STRIPS PLANNING

Quantum adiabatic hardware currently accepts problems that are in Quadratic Unconstrained Binary Optimization (QUBO) form (as discussed in Section 2.5.5). Thus, search and planning problems must be parameterized as QUBOs so that a solution can be achieved with adiabatic hardware. Rieffel et al. [2015] suggests two ways that a STRIPS instance can be mapped onto quantum adiabatic machines: (1) the Time Slice Approach and (2) the Conjunctive Normal Form (CNF) Approach.

Time Slice Approach

In the Time Splice approach, the plan length L must be known in advance. Optimization is performed over possible plans of length L that achieve all goals. An overall cost function is specified as:

$$H = H_{initial} + H_{goal} + H_{precond} + H_{effects} + H_{no-op}. \tag{3.4}$$

Equation (3.4) represents a summation of Hamiltonian terms that form an overall Hamiltonian. The detailed definition of these terms is described in Table 3.1.

Table 3.1: Time slice approach to formulating STRIPS as QUBO

Mathematical Term Definition	Term Purpose
$H_{initial} = \sum\limits_{i \in I^{(+)}} (1 - x_i^0) + \sum\limits_{i \in I^{(-)}} (x_i^0)$	Specify initial state
$H_{goal} = \sum\limits_{i \in G^{(+)}} (1 - x_i^L) + \sum\limits_{i \in G^{(-)}} x_i^L$	Specify goal state (at end of planning)
$H_{precond} = \sum\limits_{t=1}^{L} \sum\limits_{j=1}^{M} [\sum\limits_{i \in C_j^{(+)}} (1 - x_i^{t-1}) y_j^t$ $+ \sum\limits_{i \in C_j^{(-)}} x_i^{t-1} y_j^t]$	Penalize violation of preconditions
$H_{effects} = \sum\limits_{t=1}^{L} \sum\limits_{j=1}^{M} (\sum\limits_{i \in E_j^{(+)}} y_j^t (1 + x_i^{t-1} - 2x_i^t)$ $+ \sum\limits_{i \in E_j^{(-)}} y_j^t (2x_i^t - x_i^{t-1}))$	Penalize inappropriate effects
$H_{no-op} = \sum\limits_{t=1}^{L} \sum\limits_{i=1}^{N} [x_i^{t-1} + x_i^t - 2x_i^{t-1} x_i^t]$	Penalize variable changes

x_i^t represents one of L binary predicate state variables at a particular time t. x_i^t is 1 if the robot is in the i-th state at time t and 0 otherwise. y_j^t represent one of M actions in the planning problem. y_j is 1 if the robot takes action j at time t and 0 otherwise. $I^{(+)}$ represents the set of state variables that are 1 (true) in the initial state, and $I^{(-)}$ represents the set of those that are 0 (false) in the initial state. $G^{(+)}$ is the set of goal state variables with value 1, and $G^{(-)}$ is the set of those with value 0. $C_j^{(+)}$ represents the set of preconditions set to 1 for action j, while $C_j^{(-)}$ represents the set of preconditions set to 0 for that action. $E_j^{(+)}$ represents the set of effects set to 1 for action j, while $E_j^{(-)}$ represents the set of effects set to 0 for that action.

The terms in Equation (3.4) serve different purposes. $H_{initial}$ and H_{goal} encode the initial and goal states respectively, providing a penalty when the conditions of the initial and goal states are not met. $H_{precond}$ and $H_{effects}$ penalize violation of preconditions and inapplicable effects in order to choose coherent action sequences to get from the start to the goal. H_{no-op} penalizes state variable changes, helping prevent state changes that are not the result of an action to aid stability in the optimization.

The model assumes one action per time step. Additional terms can be added to allow the robot to perform multiple actions at a time step. The equations can also be simplified for specific navigation or scheduling problems. Jointly, Equation (3.4) and Table 3.1 define a QUBO that can be deployed on adiabatic hardware.

Conjunctive Normal Form (CNF) Approach

A Conjunctive Normal Form (CNF) expression is a set of clauses $\{C_a\}$. Each clause C_a consists of members from a set of n Boolean variables $\{z_i\}$ that are logically-ORed together. Different clauses can have different numbers of variables. To solve a CNF expression, all the clauses must be satisfied.

SATPLAN [Kautz and Selman, 2006] is a classical planning approach that converts a STRIPS planning problem to equivalent CNF. SATPLAN also prunes the CNF for various logical inconsistencies.

A CNF produced by SATPLAN defines a Polynomial Unconstrained Binary Optimization (PUBO) instance, a generalization of QUBO where the objective function is a pseudo-Boolean function of arbitrary degree. For each clause in the CNF, there is a term in the PUBO instance. Higher-order degree terms are iteratively replaced with ancillary variables. Variables are iteratively removed until the PUBO is quadratic. The resulting QUBO instance is deployed on quantum adiabatic hardware.

Reported Comparisons

In their experiments, Rieffel et al. [2015] find that the Time Slice and CNF approaches are relatively comparable. The generated QUBO instances are similar in size and their empirical performance on adiabatic hardware is similar, though the Time Slice is more scalable for certain situations. The quantum annealing approach shows promise, though current limitations of quantum hardware make this quantum planning approach not yet competitive with the best classical implementations of STRIPs solvers.

3.4   CHAPTER SUMMARY

<u>Chapter Key Points</u>

- Grover Tree Search attempts to extend Grover's algorithm to search in tree structures. The method suggests a quadratic speedup over classical uninformed search when the branching factor of the search tree is constant. When the branching factor is not constant, there is not always a guaranteed speedup.

- Incorporating heuristic information into quantum search is a challenge since quantum operators must be unitary. Attempts have been made to define unitary operators that incorporate heuristic information into tree search, but there is not always a guaranteed speedup.

- STRIPS is a classical framework for robotic planning. Attempts have been made to apply quantum annealing to speed up STRIPS planning, though methods are not yet competitive with the best classical STRIPS solvers.

Table 3.2: Summary of quantum operating principles in discussed quantum search methods

Technique	Quantum Operating Principles (QOPS) Used	Potential Advantages (to classical version)
Uninformed Grover Tree Search [Tarrataca and Wichert, 2011]	Grover's Algorithm	Quadratic speedup for constant branching factor
Informed Grover Tree Search [Tarrataca and Wichert, 2011]	Grover's Algorithm	Depends on Problem Instance
Quantum STRIPS [Rieffel et al., 2015]	Adiabatic Theorem	Depends on Problem Instance

CHAPTER 4

Quantum Agent Models

Agent models are often used in classical robotics to model a robot planning in a world with uncertainty. Classical environments are deterministic from time step to time step but can have a large dynamic state space that can be difficult to predict. The dynamic nature of many real-world environments means robots have to reason over a large range of possible future scenarios in planning their actions. To make matters more challenging, the robot may be uncertain of the state of the environment (as well as its own state) as it has to infer these properties from noisy, imperfect sensor data. Frameworks such as the Markov Decision Process (MDP) [Bellman, 1957] and Partially Observable Markov Decision Process (POMDP) [Kaelbling et al., 1998] have traditionally been used to model these "planning with uncertainty" scenarios.

Planning with uncertainty is a central challenge in many areas of robotics, and MDPs and POMDPs have widespread use. They have been used to model grasping and manipulation tasks [Glashan et al., 2007, Horowitz and Burdick, 2013, Hsiao et al., 2007, Pajarinen and Kyrki, 2015], robot navigation [Koenig and Simmons, 1998, Simmons and Koenig, 1995], path planning [Cortes et al., 2008, Spaan and Vlassis, 2004], target tracking [Hsu et al., 2008], high-level behavior control [Pineau and Thrun, 2002], and many other applications [Cassandra, 1998].

Often times, robots are designed with a goal in mind to solve a particular problem. As a robot performs actions, it may be able to improve its performance on its design goal over time by learning. Frameworks such as Reinforcement Learning (RL) [Sutton and Barto, 1998], Online Learning [Bottou, 1998], and Active Learning [Settles, 2010] are common paradigms for incorporating learning mechanisms into robot action, so a robot can improve using past experience.

While the field of quantum agents is still in its infancy, some work has been done to generalize MDPs and POMDPs to the quantum world and to formulate quantum speedups for agent learning. One direction of research is to develop agent algorithms that can operate in quantum-scale environments. Attempts have been made to formulate qMDPs [Ying and Ying, 2014] and QOMDPs [Barry et al., 2014] for agents attempting to operate within the unitary dynamics of a quantum environment.

Another area of active research is to develop agents that operate in classical environments but get a potential quantum speedup in their learning capability. Reinforcement Learning that uses quantum random walks for speedup has been proposed [Paparo et al., 2014]. Simple agent decision making algorithms (such as multi-armed bandits) have also been implemented in optical and photonic setups [Mahajan and Teneketzis, 2008, Naruse et al., 2013].

4.1 CLASSICAL MARKOV DECISION PROCESSES

A Markov Decision Process (MDP) [Bellman, 1957] is a model of an agent operating in an uncertain but observable world. Formally, an MDP is a 5-tuple (S, A, T, R, γ):

Markov Decision Process (MDP) (S, A, T, R, γ) Definition

- S is a set of robot-environment states.

- A is a set of actions the agent can take.

- $T(s_i, a, s_j) : S \times A \times S \rightarrow [0, 1]$ is a state transition function which gives the probability that taking action a in state s_i results in state s_j.

- $R(s_i, a) : S \times A \rightarrow \mathbb{R}$ is a reward function that quantifies the reward the agent gets for taking action a in state s_i.

- $\gamma \in [0, 1)$ is a discount factor that trades off between immediate reward and potential future reward.

In the MDP model, the world is in one state at any given time. The agent fully observes the true state of the world and itself. At each time step, the agent chooses and executes an action from its set of actions. The agent receives some reward based on the reward function for its action choice in that state. The world then probabilistically transitions to a new state according to the state transition function. The objective of the MDP agent is to act in a way that allows it to maximize its future expected reward.

The solution to an MDP is known as a policy, $\pi(s_i) : S \rightarrow A$, a function that maps states to actions. The optimal policy over an infinite horizon is one inducing the value function:

$$V^*(s_i) = \max_{a \in A} \left[R(s_i, a) + \gamma \sum_{s_j \in S} T(s_i, a, s_j) V^*(s_j) \right]. \tag{4.1}$$

This famous equation, known as the "Bellman Equation," quantifies the value of being in a state based on immediate reward as well as discounted expected future reward. The future reward is based on what the agent can do from its next state and the reward of future states it may transition to, discounted more as time increases. The recursion specifies a dynamic program which breaks the optimization into subproblems that are solved sequentially in a recurrent fashion.

The Bellman Equation is amenable to solutions via the Value Iteration algorithm [Bellman, 1957]. In general, there is a unique solution for V^* that is non-infinite if $\gamma < 1$. When the input size is polynomial in $|S|$ and $|A|$, finding an ϵ-optimal policy (i.e., one within ϵ cost of the true optimal policy) for an MDP can be done in polynomial time.

4.2   CLASSICAL PARTIALLY OBSERVABLE MARKOV DECISION PROCESSES

Partially Observable Markov Decision Processes (POMDPs) [Russell and Norvig, 2003] generalize MDPs to the case where the world is not fully observable. In this setting, the true world state is hidden to the agent, though the agent still receives observations of the world that help it infer the world state. POMDPs have been useful for modeling robots because imperfect robot sensors and actuators provide only an inherently limited view of the world.

Formally, a POMDP is an 8-tuple $(S, A, T, R, \gamma, \Omega, O, \vec{b_0})$. It has all 5 components of an MDP and has in addition:

Partially Observable Markov Decision Process (POMDP) $(S, A, T, R, \gamma, \Omega, O, \vec{b_0})$ Definition

- Ω, a set of possible observations of the world.

- $O(s_i, a_j, o_k) : S \times A \times \Omega \to [0, 1]$, an observation model indicating the probability of making observation o_k given that action a_j was taken and the system ended up in state s_i.

- $(\vec{b})_0$, a probability distribution over possible initial states.

The world in the POMDP, like the MDP, functions deterministically. However, unlike the MDP, an agent never knows exactly the state in which it resides. Instead, the agent maintains a probability distribution over possible states and attempts to statistically infer the system state from the observation data it receives. In a POMDP, the agent maintains, at each time step, a vector $\vec{b}$, a probability distribution (called the "belief state") over possible states. For $s_i \in S$, $\vec{b_i}$ is the probability the system is in state s_i. Thus, $0 \le \vec{b_i} \le 1$ and $\sum_i \vec{b_i} = 1$.

Interestingly, POMDPs can be viewed as MDPs in belief space [Kaelbling et al., 1998]. Given a POMDP, one can define a corresponding "belief MDP" where the state space is defined as the set of all possible belief states of the POMDP. The optimal solution to the belief MDP is also the optimal solution to the POMDP. This observation doesn't make finding a solution easier since the state space of the belief state MDP is continuous and all known algorithms for solving MDPs optimally in polynomial time are polynomial in the size of the state space.

As with MDPs, the solution for a POMDP is a policy which dictates what action to take in any state of the robot-environment system. While efforts have been made to solve POMDPs using a myriad of techniques [Braziunas, 2003, Murphy, 2000, Roy et al., 2005, Shani et al., 2013], classical POMDPs do not have optimistic theoretical properties. The policy existence problem for

POMDPs is PSPACE-hard [Papadimitriou and Tsitsiklis, 1987] and undecidable in the infinite horizon [Madani et al., 2003].

4.3   QUANTUM SUPEROPERATORS

Efforts have been made to generalize MDPs to quantum MDPs (qMDPs) [Ying and Ying, 2014] and POMDPs to quantum POMDPs (QOMDPs) [Barry et al., 2014] to model control systems operating on quantum-scale environments. In both frameworks, quantum superoperators formalize an agent's capability to affect its quantum world and the expected quantum state evolution that results from the action.

A quantum superoperator is a linear operator acting on a vector space of linear operators. In the quantum world, superoperators can be used to evolve density matrices of quantum states. Formally, a quantum superoperator $S = \{K_1, \ldots, K_\kappa\}$ acting on states of dimension d is defined by a set of κ $d \times d$ Kraus matrices. A set of matrices is a set of Kraus matrices if it satisfies $\sum_{i=1}^{\kappa} K_i^\dagger K_i = I_d$.

When a quantum superoperator operates on a density matrix ρ, there are κ possible next states for ρ. The next state ρ_i is sampled from the probability distribution of possible next states. The next state is given by $\rho_i' \rightarrow \frac{K_i \rho K_i^\dagger}{Tr(K_i \rho K_i^\dagger)}$ with probability $P(\rho_i'|\rho) = Tr(K_i \rho K_i^\dagger)$. The superoperator returns observation i if the ith Kraus matrix is applied.

4.4   QUANTUM MDPS

A quantum MDP (qMDP) [Ying and Ying, 2014] is a 4-tuple (H, Act, M, Q):

Quantum MDP (qMDP) (H, Act, M, Q) Definition

- H is the set of possible quantum states of the system. In quantum mechanics, H is a Hilbert space, and states in H are density matrices that can be either mixed or pure states.

- Act is a finite set of actions. Each $\alpha \in Act$ has a corresponding superoperator ε_α that can be used to evolve the system.

- $\mathbb{M}$ is a finite set of quantum measurements. The set of all possible observations is $\Omega = \{O_{M,m} : M \in \mathbb{M}$ and m is a possible outcome of $M\}$. $O_{M,m}$ indicates we perform the measurement M and obtain the outcome m.

- $Q : Act \cup \mathbb{M} \rightarrow 2^{Act \cup M}$ is a mapping from current measurements and actions to the set of available measurements and actions after a certain current action or current measurement is performed.

For the qMDP, the world is quantum-scale and thus governed by unitary dynamics. The possible states of the quantum system are described by the Hilbert space H. The quantum state is a possible superposition of states, with the true underlying state being unknown. At each time step, the agent can choose to perform an action or a measurement (to try to deduce the underlying state).

With a qMDP, a series of measurements is introduced to infer the state and help select the next action, based on the outcomes of said measurements. Obtaining information about a quantum system in superposition without decoherence is non-trivial. One method is to use indirect measurements where the main quantum system interacts with an ancilla quantum system. The coupled ancilla system is subject to measurement to obtain indirect information about the underlying system, though the main system remains unmeasured. This measurement methodology is further described in Section 6.1.2.

The algorithm that utilizes the obtained measurement information for decision making is called a scheduler. A scheduler for a qMDP is a function $\mathfrak{S} : (Act \cup \Omega)^* \to Act \cup \mathbb{M}$ that selects the next action based on the outcomes of previously performed measurements and actions. A key difference between MDPs and qMDPs is that with MDPs, the scheduler can use the exact state of the system because the state of the classical world is known in the MDP. In a quantum environment, the true world state is not known and must be inferred from partial measurements of the system that give information about the system but do not collapse the superposition.

Classically, for an MDP, it can be shown that a memory-less scheduler exists that is optimal for all initial states [Baier and Katoen, 2008]. In the quantum case, however, there is no guarantee that an optimal scheduler exists for even a fixed initial state. The challenge lies in knowing the current state of the quantum system. A quantum system in a superposition of multiple states does not have a definite current state and thus does not provide enough information for a scheduler to choose a next best action.

Even if one knows the superposition weights and possible states, the uncertain knowledge of the true state makes the qMDP a very different mathematical object from the classical MDP. The reachability problem in MDPs is to determine whether a given scheduler will reach a particular state. Ying and Ying [2014] provides some results for the reachability properties of qMDPs. For the finite horizon case, the reachability problem for a qMDP is undecidable. For the infinite horizon case, it is EXPTIME-hard to determine whether the supremum reachability probability of a qMDP is 1.

An invariant subspace is a subspace of H that is invariant under all superoperators ε_α for all $\alpha \in Act$ and invariant under all measurements $M \in \mathbb{M}$. An invariant subspace can thus be viewed as a type of stationary region for a qMDP. It is proven by Ying and Ying [2014] that a qMDP has an optimal scheduler for reaching an invariant subspace of its Hilbert space H if and only if the orthocomplement of the target subspace contains no invariant subspaces. Using this property, Ying and Ying [2014] develops an algorithm for finding an optimal scheduler for a qMDP, when one exists.

4.5   QOMDPS

As discussed in the previous section, the qMDP is dependent on measurements to infer environment state, something that is not necessary in classical MDPs. When measurements are needed to infer environment state, the environment is classically, by definition, partially observable. The partial observability of the true underlying state of a quantum system in superposition may suggest POMDPs as a potentially better model of quantum systems than MDPs.

A quantum POMDP (QOMDP) [Barry et al., 2014] is a 6-tuple $(S, \Omega, \mathbb{A}, R, \gamma, \rho_0)$:

Quantum POMDP (QOMDP) $(S, \Omega, \mathbb{A}, R, \gamma, \rho_0)$ Definition

- S, a set of possible quantum states of the system. In quantum mechanics, S is a Hilbert space and states in S are density matrices that can be either mixed or pure states.

- $\Omega = \{o_1, \ldots, o_{|\Omega|}\}$, a set of possible observations.

- $\mathbb{A} = \{A^1, \ldots, A^{|\mathbb{A}|}\}$, a set of superoperators that allow the agent to manipulate the quantum system. Each superoperator $A^a = \{A_1^a, \ldots, A_{|\Omega|}^a\}$ has $|\Omega|$ Kraus matrices. Each superoperator returns the same set of possible observations. The return of o_i means the ith Kraus matrix was applied. Taking action a in state ρ returns observation o_i with probability $P(o_i | \rho, a) = Tr(A_i^a \rho A_i^{a\dagger})$. If o_i is observed after taking action a in state ρ, the next state is $N(\rho, a, o_i) = \frac{A_i^a \rho A_i^{a\dagger}}{Tr(A_i^a \rho A_i^{a\dagger})}$.

- $R = \{R_1, \ldots, R_{|A|}\}$, is a set of operators for receiving the reward. The reward associated with taking action a in state ρ is $R(\rho, a) = Tr(\rho R_a)$.

- $\gamma \in [0, 1)$, the discount factor.

- $\rho_0 \in S$, the starting state.

The agent workflow in QOMDPs is analogous to that of classical POMDPs. For a QOMDP, the world is a quantum system that begins in state ρ_0 and then evolves based on the agent's actions.

Quantum superoperators formalize an agent's capability to take an action in a quantum environment and receive a percept. In the QOMDP, acting and sensing in the environment are coupled via superoperators. At each time step, the agent chooses a superoperator to apply from its set of superoperators. Observation of the system may be done (often via indirect measurement methodologies), and the agent receives an observation from Ω according to the laws of quantum mechanics. The agent also receives a reward given by R based on the state of the system after the superoperator is applied.

In a classical POMDP, the environment is always in one underlying state which is hidden from the agent. The POMDP agent has a belief space over the states of the environment. In a QOMDP, the true underlying state may be a superposition. The QOMDP agent knows the state probabilities of the superposition because of $N(\rho, a, o_i)$.

As with POMDPs, a policy for a QOMDP is a function $\pi : S \rightarrow A$, mapping of states to actions. The value of the policy is:

$$V^*(\rho_0) = R(\rho_0, \pi(\rho_0)) + \gamma \sum_{i=1}^{|\Omega|} P(o_i|\rho_0, \pi(\rho_0))V^\pi(N(\rho_0, \pi(\rho_0), o_i)). \qquad (4.2)$$

Equation 4.2 is the QOMDP version of the classical Bellman equation. However, POMDPs and QOMDPs are quite different mathematical objects with different theoretical properties. Because classical POMDPs can be simulated within the framework of the QOMDP, the QOMDP inherits the hardness properties of the POMDP [Barry et al., 2014].

The policy existence problem is to decide, given a decision process D, a starting state s, horizon h, and value V, whether there is a policy of horizon h that achieves value at least V for s in D. If h is infinite, the policy existence problem is undecidable for both POMDPs and QOMDPs [Barry et al., 2014]. If h is polynomial in the size of the input, the policy existence problem is in PSPACE [Barry et al., 2014].

The goal-state reachability problem of POMDPs is, given a goal decision process D, starting state s, determine whether there exists a policy that can reach the goal state from s in a finite number of steps with probability 1. For POMDPs, the goal-state reachability problem is decidable. However, for QOMDPs, Barry et al. [2014] proves that goal-state reachability is undecidable.

4.6   CLASSICAL REINFORCEMENT LEARNING MODELS

Classical Reinforcement Learning (RL) in robotics is used to incorporate learning capability into the planning mechanisms of MDPs and POMDPs. RL models an embodied agent situated and operating in a dynamic environment. The robotic agent receives sensory percepts to understand the current world state and has mechanisms for performing actions to affect the physical environment. The robotic agent typically has an objective or goal to accomplish and memory to allow for learning to improve its performance over time with respect to the goal.

Key to the RL model is that the environment (or other feedback mechanism) provides some kind of reward signal after the agent acts to help it learn. The agent uses this reward as a feedback signal, helping it know whether it is performing well or poorly in accomplishing its task. Each time step for the agent consists of a percept-action-reward cycle. After each time step, the agent learns from experience to update its internal processing to perform better in the future. The agent can, over time, learn which actions to use in what states to maximize performance with respect to its objective.

A simple RL agent can be described by a 6-tuple (S, A, Λ, I, D, U):

Reinforcement Learning Agent (S, A, Λ, I, D, U) Definition

- $S = \{s_1, s_2, \ldots, s_m\}$, a set of possible environment percepts

- $A = \{a_1, a_2, \ldots, a_n\}$, a set of possible actions

- $\Lambda = \{0, 1\}$, a reward set which describes environment reward. This is the very simple case of reward being either "0" for negative reinforcement and "1" for positive reinforcement.

- $I = \{i_1, i_2, \ldots, i_p\}$, a set of internal states of the agent.

- $D : S \times I \to A$, a function which maps percepts and internal state to action.

- $U : S \times A \times \Lambda \times I \to I$, a function which updates the internal state via the reward of the last percept-action-reward sequence.

The mathematical framework is a very simple template for a reinforcement learning agent, of which there are many concrete variations. Some classes of reinforcement learning agents include Projection Simulation (PS) and Reflective Projection Simulation (RPS) agents [Briegel and De las Cuevas, 2012].

4.6.1   PROJECTION SIMULATION AGENTS

Projection Simulation (PS) agents are agents whose internal states represent Episodic and Compositional Memory (ECM), consisting of directed, weighted graphs. The vertices of the graphs are "clips" that represent fragments of episodic experience (internal representation of received percepts and possible actions). Edges in the graphs represent the transition probability between clips of episodic experience. Collectively, the graphical structures represent the agent's "brain."

Deliberation consists of association-driven hops between the memory sequence clips in the graphs. The elementary process can be described by a Markov chain with the clips (i.e., graph vertices) and transition probabilities (i.e., graph edge weights). Given a new percept, the corresponding clip in the ECM is excited, and hopping between clips occurs according to the specified transition probabilities. In the simplest model, the hopping process commences once a unit-length action clip is encountered. This unit-length action clip is then sent to the robot actuators. The ECM thus functions as a type of associative memory where, when the agent receives similar percepts, it traverses similar memory clips in the graphical model and acts according to the best policy it has learned thus far.

Learning involves changing the edge weights of the graph to upweight or downweight the probability of transitioning to particular clips (i.e., percepts and/or action sequences). Learning in the PS agent occurs according to the following update rules for h, the matrix containing the edge weights of the graph. At each time step the PS agent receives a percept, takes an action, and receives a reinforcement signal. If the action the agent takes is rewarded positively with $r_t \in \Lambda$ and clips c_i and c_j are traversed in the hopping process to output that action, the connection between the clips is strengthened:

$$h^{(t+1)}(c_i, c_j) = h^{(t)}(c_i, c_j) - \gamma \left[h^{(t)}(c_i, c_j) - 1 \right] + r_t \tag{4.3}$$

where $r_t > 0$ is a positive reward and $0 \leq \gamma \leq 1$ is a "forgetfulness" parameter that trades off between the current reward and the previously learned model. If the action was not rewarded positively or clips c_i and c_j were not played in the hopping process:

$$h^{(t+1)}(c_i, c_j) = h^{(t)}(c_i, c_j) - \gamma \left[h^{(t)}(c_i, c_j) - 1 \right]. \tag{4.4}$$

The learning rules strengthen clip connections that help obtain positive reward while downweighting ones that are detracting from or are not useful toward obtaining positive reinforcement. The weights in h are normalized to form probabilities after each update.

With a large ECM, traversing edges between clips to get to an action can be slow. To speed up learning, the PS agent also maintains $F = \{f(s) | f(s) \subseteq A, s \in S\}$, a non-empty set of flagged actions for each percept. This set represents the agent's short term memory, and is used to speed up learning. If, given a percept, an action is taken but not rewarded, the action is removed from the set of flagged actions.

4.6.2   REFLECTIVE PROJECTION SIMULATION AGENTS

Reflective Projection Simulation (RPS) agents are PS agents who repeat their diffusion process many times to attempt to find the stationary distribution of their Markov chains. They approximate the complete mixing of their Markov chains, simulating infinite deliberation times, as opposed to just the finite deliberation of PS agents. Once the mixing converges, an RPS agent samples from the stationary distribution over the clip space until an action clip is reached.

To implement this design, RPS agents utilize a random walk algorithm on Markov chains. Formally, a Markov chain is specified by a transition probability matrix P with entries P_{ji}, the probability of moving from clip j to clip i in the graph. For an irreducible Markov chain, there exists a stationary distribution π such that $P\pi = \pi$. For an irreducible and aperiodic Markov chain, one can approximate this distribution by $P^t \pi_0$ (i.e., applying P to the initial distribution π_0 to simulate the random state evolution after t time steps) where $t \geq t_{\epsilon'}^{\mathrm{mix}}$.

The mixing time, $t_{\epsilon'}^{\mathrm{mix}}$, satisfies:

$$\frac{1}{\delta} \frac{\lambda_2}{\log 2\epsilon'} \leq t_{\epsilon'}^{\mathrm{mix}} \leq \frac{1}{\delta} \left[\max_i (\log \pi_i^{-1}) + \log (\epsilon')^{-1} \right] \tag{4.5}$$

where δ is the spectral gap of the Markov chain, defined as $\delta = 1 - |\lambda_2|$ (λ_2 is the second largest eigenvalue of P), and ϵ' is the probability of sampling a particular action from the stationary distribution.

The internal state of an RPS agent is an irreducible, aperiodic, and reversible Markov chain over subsets of the clip space. For a percept s, let π_s be the stationary distribution of the current Markov chain P_s and $f(s)$ be the flagged action (i.e., short term memory) function which maps percepts to actions. The RPS computes the stationary distribution of P_s as $\pi_s(i)$. The RPS agent outputs an action according to:

$$\tilde{\pi}_s(i) = \begin{cases} \dfrac{\pi_s(i)}{\sum\limits_{j \in f(s)} \pi_s(j)} & \text{if } i \in f(s) \\ 0 & \text{otherwise} \end{cases} \tag{4.6}$$

which is the renormalized stationary distribution with support constrained to only flagged actions. The agent prepares this probability distribution and samples clips using it until a flagged action is hit.

The RPS, by approximating the complete mixed distribution, simulates infinite deliberation time. The main computational requirement is running the algorithm for enough iterations to produce a convergent sampling distribution. This is where agent learning deployed on quantum computers may provide a speedup compared to classical implementations of RPS agents.

4.7   QUANTUM AGENT LEARNING

The model of quantum agent learning in Paparo et al. [2014] preserves the classical robot learning context. The robotic agent is operating in a physical environment that is, for practical purposes, entirely classical. The robotic agent thus cannot query the environment or try actions in quantum superposition since that would violate the laws of classical physics. The quantum RPS agent, however, can potentially obtain a speedup in its internal deliberation processes. By being able to deal with larger problem instances, a quantum robotic agent could theoretically operate in more unstructured and complicated environments than classical robotic agents.

The quantum RPS agent leverages search via quantum random walks on Markov chains [Magniez et al., 2011, Szegedy, 2004]. The quantum RPS agent uses several standard quantum operators such as the quantum diffusion operator U_{P_s} and quantum-walk operator to construct two key pieces of mathematical machinery: the approximate reflection operator subroutine $R(P_s)(q, k)$ and the reflection onto the set of flagged actions $ref[f(s)]$. The quantum active learning algorithm (shown in Algorithm 4.1) is essentially an application of these two components.

The approximate reflection operator subroutine $R(P_s)(q, k)$ (as its name suggests) approximates the reflection operator $2 |\pi_s\rangle \langle \pi_s| - 1$, where $|\pi_s\rangle = \sum_i \sqrt{\pi_s(i)} |i\rangle$ is the encoding of the stationary distribution π_s. The parameters q and k control how accurate the approximation is and how many times the quantum-walk operator needs to be applied. The distance between the

Algorithm 4.1 Quantum Active Learning Algorithm

Input : Initial state
Output : Flagged action to take

1: Initialize register to quantum state $|\pi_{init}\rangle = U_{P_s} |\pi_s\rangle |0\rangle$
2: Agent performs sequence of reflections over flagged actions $ref[f(s)]$ interlaced with applications of the approximate reflection operator $R(P_s)(q, k)$.
3: Resulting quantum state measured and the found flagged action is checked. If the output clip is not an action, go back to step 1.
4: **return** Flagged action

approximate subroutine and ideal reflection operator is upper bounded by 2^{1-k} under a suitable metric. The fidelity of operator approaches unity exponentially quickly in k, making it a good approximation. $ref[f(s)]$ is the operator for reflections over flagged actions. This is a Grover-like reflection which, after it has been applied some number of times, finds a flagged action upon measurement. The operator is used with the search algorithm described in Magniez et al. [2011].

The probability of not hitting a flagged action decreases exponentially with the number of algorithm iterations. To find the stationary distribution, the total number of calls to quantum diffusion operators is $O(\frac{1}{\sqrt{\epsilon_s \delta_s}})$ where $\delta_s = 1 - |\lambda_2|$ (the "spectral gap" of P_s), λ_2 is the second largest eigenvalue of P_s, and $\epsilon_s = \sum_{i \in f(s)} \pi_s(i)$ (the probability of sampling a flagged action from stationary distribution π_s). The total required number of reflections over flagged actions is $O(\frac{1}{\sqrt{\epsilon_s}})$.

A classical agent and a quantum agent who perceive the same percepts and can produce the same actions are proved to be $\epsilon-$equal in terms of produced behavior [Paparo et al., 2014]. The behavioral equivalence of classical and quantum RPS agents means the solutions they learn are not intrinsically different. However, the quantum robotic agent is significantly faster in finding the solution, producing, in theory, a quadratic speedup over the classical robotic agent [Paparo et al., 2014].

Some mechanisms for implementing quantum RPS agents are suggested by Paparo et al. [2014]. Quantum random walks and related processes can be naturally implemented in linear optics setups by, for instance, arrays of polarizing beam splitters [Aspuru-Guzik and Walther, 2012]. Internal states of trapped ions [Blatt and Roos, 2012] could also be used. Finally, condensed matter systems in which the proposed Markov chains could be realized through cooling or relaxation processes toward target distributions that encode the state of belief of the agent are also a potential implementation strategy [Schaetz et al., 2013].

4.8   MULTI-ARMED BANDIT PROBLEM AND SINGLE PHOTON DECISION MAKER

The multi-armed bandit problem (defined in Mahajan and Teneketzis [2008]) describes the task of a gambler who has the choice of spinning 1 of N slot machines at each time step. The slot machines differ in their expected payoffs, which are unknown to the gambler *a priori*.

To improve profits over time, the gambler can learn from the experience of winning and losing from playing the slot machines. A slot machine that contributes to the gambler's total gain is a natural choice for the next run, though the gambler must identify those favorable slot machines before he/she can exploit them. Thus, the complexity of optimally sequentially playing the slot machines is associated with an initial "exploration" phase where the gambler tries different slot machines to learn their distribution (which may lose some reward) and a subsequent "exploitation" phase where the most profitable slot machines are played repeatedly to maximize profit.

The multi-armed bandit problem is an analogy for many decision-making settings where there are many options but the expected rewards are uncertain. In robotics, multi-armed bandit problem formulations have been proposed for exploring unknown environments [Si et al., 2007], robot foraging [Srivastava et al., 2013], and task partitioning in swarm robotics [Pini et al., 2012]. There are many classical approaches to tackling these problems, but here we discuss an exciting quantum implementation.

Efforts have been made to solve multi-armed bandit problems using photonic implementations. For the $N = 2$ case of the multi-armed bandits problem, [Naruse et al., 2015] demonstrates a physical implementation that uses a single photon as a decision maker. The system implementation diagram is shown in Figure 4.1.

In the implementation, a laser excites a nitrogen-vacancy in a diamond crystal, and the emitted single photon passes through an immersion objective (IO), polarization adjuster apparatus (PAA), and a polarization beam splitter (PBS). The played slot machine at each time step (either the right or left slot machine) depends on the detection of a single photon at the reflected or transmitted beam of the PBS respectively. A winner-loser detector closes the feedback loop by adjusting the PAA to select slot machines that increase the total gain or suppress the choice of slot machine that did not reward the gambler.

The proposed physical implementation of multi-armed bandit decision making in optics benefits from the quantum nature of the photon. Decision making could feasibly occur at the speed of light without additional computational effort. Note, however, that, at the moment, the entire optical system has additional latencies and can take several seconds to read out a decision. The current implementation is thus not yet a real-time decision-making system. In addition, much work remains to be done to scale the system to larger than $N = 2$ in the multi-armed bandits problem.

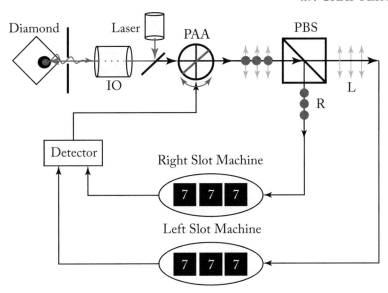

Figure 4.1: Illustration of single photon decision maker. (Adapted from "Single-photon decision maker," by Naruse et al. [2015], Scientific Reports 5: 13253. Adapted with permission.)

4.9   CHAPTER SUMMARY

<u>Chapter Key Points</u>

- Markov Decision Processes (MDPs) and Partially Observable Markov Decision Processes (POMDPs) are classical frameworks for modeling robotic planning with uncertainty.

- Quantum MDPs (qMDPs) and Quantum POMDPs (QOMDPs) attempt to model an agent tasked with manipulating quantum phenomena in a quantum environment. The models have different complexity properties from their classical counterparts.

- Reinforcement Learning is a classical framework for modeling a robot that learns to improve its performance in a dynamic environment with feedback from a teacher.

- Attempts have been made to formulate Quantum Agent Learning algorithms that may help a robot making decisions in a classical environment improve its behavior over time. In theory, a quantum agent may be able to learn faster than a classical agent to reach optimal behavior.

- Attempts have been made to solve simple multi-armed bandit instances with a photon decision maker.

Table 4.1: Summary of quantum operating principles in discussed quantum agent models

Technique	Quantum Operating Principles (QOPS) Used	Potential Advantages (to classical version)
qMDP [Ying and Ying 2014]	Quantum Superoperators	Planning capability in quantum environments
QOMDP [Barry et al., 2014]	Quantum Superoperators	Planning capability in quantum environments
Quantum Active Learning [Paparo et al., 2014]	Quantum Random Walks	Quadratic speedup for reinforcement learning in classical environments
Photon Decision Maker [Naruse et al., 2015]	Photonic Hardware	Faster decision making in classical environments

CHAPTER 5

Machine Learning Mechanisms for Quantum Robotics

Applications of machine learning are ubiquitous in robotics today. Agent learning (previously covered) is only one of the forms of machine learning commonly used in classical robotics. We briefly summarize a few areas of classical robotics that have recently used machine learning:

- Automated analysis and data mining of sensor data for perception: A robot needs to be able to robustly understand its sensor data, mine it for patterns, and understand possible percepts (e.g., obstacles) in its world. Machine learning approaches are commonly being researched for computer vision [Cipolla et al., 2013], lidar [Lodha et al., 2006, Zhao et al., 2011], sonar [Dietterich, 1997], force sensing [Edsinger-Gonzales and Weber, 2004], and other such perception technologies.

- Robot localization: Robots need to be able to self-localize in an environment. Modern localization technologies (e.g., inertial navigation [Bagnell et al., 2010], wi-fi localization [Biswas and Veloso, 2010], range-sensor-based localization [Thrun et al., 2001]) all make extensive use of machine learning methods.

- Learning robot controllers and planners: Learning robot controllers from simulation or actual data are key techniques being employed in many domains involving robots. High-level robot planners also benefit from learning, such as learning retrospectively from experience. Online Learning [Gaskett and Cheng, 2003, Hadsell et al., 2007, Hagras, 2001], Reinforcement Learning [Kaelbling et al., 1996], and Deep Learning [Bengio, 2009, LeCun et al., 2010, Lenz et al., 2015] are key approaches being heavily researched by groups all over the world.

- Learning human-robot interaction: Data-driven approaches are being used to understand how humans and robots interact [Hiraki, 1995]. Machine learning approaches are being used to model this interaction and recognize user affect [Rani et al., 2006].

Given the extensive and varied ways that classical machine learning is currently being used in robotics, one expects that fundamental advances in machine learning will be impactful. Quantum machine learning is a rapidly emerging research discipline. For excellent surveys of the quantum machine learning literature, we direct the reader to Wittek [2014] and Schuld et al. [2015]. In our exposition in quantum robotics, we are indebted to the literature summaries provided by these and strive to build upon their work by also incorporating research released after their publication. One of our further contributions to the literature is in presenting quantum algorithms in comparison to classical robotic algorithms and highlighting the ways quantum robotics allows a speedup, a representation advantage, or, in some cases, a fundamental departure from classical robotic methods.

5.1   QUANTUM OPERATING PRINCIPLES IN QUANTUM MACHINE LEARNING

Quantum machine learning algorithms can be fairly nuanced in their mathematics. To make the concepts easier to understand, we boil down these algorithms to their key Quantum Operating Principles (QOPs) in this section. Recall from Section 1.3, QOPs is a presentation style we introduce to help explain the basic principles by which many quantum machine learning algorithms obtain benefits over classical approaches.

5.1.1   QUANTUM MEMORY

The first key QOP is the use of qubit representation in a quantum memory. By employing qubits in superposition, quantum computers are expected to have substantial representational power gains over classical methods of representing data. While N classical bits can only represent one N-bit number, N qubits can represent all possible 2^N N-bit numbers.

This capability suggests a future world of "Big Quantum Data," a term coined by Lloyd et al. [2013]. The International Data Corporation (IDC) estimates that by 2020, humans will have produced 40 zettabytes of electronic data [Gantz and Reinsel, 2012]. This entire data set could fit in about 79 qubits. It is estimated that there are 10^{90} particles in the universe. This number of possibilities can be represented with about 300 qubits. To put things into perspective, it is estimated that the human brain stores only 2.5 petabytes of binary data [Juels and Wong, 2014]. This can be stored in 55 qubits.

There is asymptotically modest overhead in converting between classical and quantum data. Classical data represented as N-dimensional complex vectors can be mapped onto quantum states over $\log_2(N)$ qubits. Description of quantum memory (QRAM) is given in Lloyd et al. [2013]. The mapping takes $O(\log_2(N))$ steps, but, once in quantum form, quantum complexities apply for algorithms. Using Grover's algorithm or the Adiabatic Theorem, many classical algorithms are estimated to obtain exponential or quadratic speedups.

5.1.2   QUANTUM INNER PRODUCTS AND DISTANCES

Quantum computing theorizes faster computation of the inner product between two vectors [Schuld et al., 2015]. Since many machine learning algorithms make use of inner products, these algorithms are expected to benefit from quantum approaches.

Lloyd's algorithm [Lloyd et al., 2013] for computing quantum inner products makes use of the polarization identity for dot products:

$$x_i^T x_j = \frac{||x_i||^2 + ||x_j||^2 - ||x_i - x_j||^2}{2}. \tag{5.1}$$

Define $Z = ||x_i||^2 + ||x_j||^2$ so $x_i^T x_j = \frac{Z - ||x_i - x_j||^2}{2}$. The first step of the algorithm involves generating the quantum states $|\psi\rangle$ and $|\phi\rangle$:

$$|\psi\rangle = \frac{1}{\sqrt{2}}(|0\rangle |x_i\rangle + |1\rangle |x_j\rangle)$$
$$|\phi\rangle = \frac{1}{Z}(||x_i|| |0\rangle - ||x_j|| |1\rangle) \tag{5.2}$$

$|\psi\rangle$ can be generated by addressing the QRAM, while the other state can be estimated using the subsequent steps. By evolving the state $\frac{1}{\sqrt{2}}(|0\rangle - |1\rangle) \otimes |0\rangle$ with the Hamiltonian $H = (||x_i|| |0\rangle \langle 0| + ||x_j|| |1\rangle \langle 1|) \otimes \sigma_x$, (where $\sigma_x = \begin{bmatrix} 0 & 1 \\ 1 & 0 \end{bmatrix}$, a Pauli matrix), the following state is obtained:

$$\frac{1}{\sqrt{2}}\Big[\cos(||x_i|| t) |0\rangle - \cos(||x_j|| t) |1\rangle \Big] \otimes |0\rangle$$
$$- \frac{i}{\sqrt{2}}\Big[\sin(||x_i|| t) |0\rangle - \sin(||x_j|| t) |1\rangle \Big] \otimes |1\rangle. \tag{5.3}$$

The parameter t is chosen subject to the constraint that $||x_i|| << 1$ and $||x_j|| << 1$. By measuring the result, the state $|\phi\rangle$ is returned with probability $\frac{Z^2 t^2}{2}$. Repeating the experiment multiple times allows estimation of Z. If the desired accuracy of estimation is ϵ, the complexity of constructing $|\phi\rangle$ and Z is $O(\epsilon^{-1})$. After a swap operation is applied, the resulting state can be used to obtain an estimate of the dot product:

$$\frac{1}{2}\Big[|0\rangle (|\psi\rangle |\phi\rangle + |\phi\rangle |\psi\rangle) + |1\rangle (|\psi\rangle |\phi\rangle - |\phi\rangle |\psi\rangle) \Big]. \tag{5.4}$$

With QRAM accesses, the overall complexity of a single dot product of an N-dimensional vector with Lloyd's dot product algorithm is $O(\epsilon^{-1} \log N)$. Classically, computing a dot product is $O(N)$.

The quantum dot product algorithm generalizes to nonlinear metrics. Harrow et al. [2009] shows that an arbitrary distance function can be approximated by q-th order polynomials. While classical non-linearity can be exponentially hard, the expected time complexity for this approximation on quantum computers is $O(\epsilon^{-1} q \log(N))$.

5.1.3   HAMILTONIAN SIMULATION

Machine learning algorithms that can be cast within a Hamiltonian framework can be simulated on a quantum computer [Berry et al., 2007]. The Lie Product Formula states that for arbitrary $n \times n$ matrices X and Y,

$$e^{X+Y} = \lim_{n \to \infty} \left(e^{\frac{X}{n}} e^{\frac{Y}{n}} \right)^n . \tag{5.5}$$

To simulate a Hamiltonian H that is a sum of m component Hamiltonians H_j:

$$H = \sum_{j=1}^{m} H_j \tag{5.6}$$

one can calculate:

$$e^{iHt} \approx \left(e^{\frac{iH_1 t}{n}} e^{\frac{iH_2 t}{n}} \ldots e^{\frac{iH_m t}{n}} \right)^n \tag{5.7}$$

where $i = \sqrt{-1}$. If the desired accuracy of the simulation is ϵ, it will be bounded by:

$$\epsilon \approx n \left(\sum_{j=1}^{m} d_j^2 \right) \leq nmd^2$$
$$d = \max_j d_j \tag{5.8}$$

where d_j is the numeric dimensionality of the Hilbert space that Hamiltonian H_j acts on. The time complexity of Hamiltonian Simulation is linear in the number of Hamiltonian steps (i.e., number of exponentials).

5.1.4   QOPS SUMMARY FOR QUANTUM MACHINE LEARNING

Table 5.1 highlights the Quantum Operating Principles (QOPs) used by specific machine learning algorithms. In the subsequent sections, we survey each of these algorithms in the context of classical robotics and how the algorithm is expected to change in the quantum world. We also hypothesize the application of the quantum algorithm to key computational problems in robotics.

5.2   QUANTUM PRINCIPAL COMPONENT ANALYSIS (PCA)

Principal Component Analysis (PCA) constructs a set of features (i.e., "principal components"), which are linear combinations of the original features, that minimize the reconstruction error of the original data. PCA has two major uses: first, as a method of data exploration for identifying the key linear directions of variance in a data set and, second, as an approach to dimensionality reduction. Generally, keeping fewer of the new (principal) components is sufficient for low reconstruction error on the original data, facilitating reduction in dimensionality of the data and, at the same time, explanation of its variance.

5.2.1   CLASSICAL PCA ANALYSIS

The input to PCA is a data matrix X (N data points $\times$ D dimensions per point). Classical PCA calculates the eigendecomposition of $X^T X$ using Singular Value Decomposition (SVD) as $X^T X = W \Sigma W^T$ where W is a matrix of eigenvectors and Σ is a diagonal matrix containing the eigenvalues in decreasing order. The projection of the data onto its top k eigenvectors or "principal components" can be calculated via $T_k = XW_k$. When the data is unprojected and the reconstruction error computed via $R = ||X - T_k W_k^T||_2$, high reconstruction error is indicative of anomalous behavior to those extracted linear directions of maximal variance. The method can thus help identify the key linear directions of variance in data and distinguish data points that are anomalous to general trends.

5.2.2   QUANTUM PCA ANALYSIS

Quantum PCA [Lloyd et al., 2014] potentially offers an exponential factor speedup when compared to classical PCA. The method computes the eigendecomposition of the data matrix using a quantum state built via simulating Hamiltonians. The key insight is that a density matrix ρ can be used as a Hamiltonian on another density matrix. n copies of the density matrix ρ are used to perform the unitary transform $e^{i\rho t}$ where $i = \sqrt{-1}$. Applying a quantum eigenvalue estimation on $e^{i\rho t}$ yields the following quantum state:

$$\sum_k r_k \, |\chi_k\rangle \, \langle \chi_k| \otimes |\hat{r}_k\rangle \, \langle \hat{r}_k| \tag{5.9}$$

where χ_k are the eigenvectors of ρ and $\hat{r}_k$ are its eigenvalues. Sampling from this state reveals the key principal components [Lloyd et al., 2014]. Classical PCA takes time polynomial in the dimension of system to compute the principal components, while Quantum PCA takes time logarithmic in the dimension of the system.

Quantum PCA also has an interesting interpretation as a method for "quantum self-tomography." In conventional sensing with robots, the environment is passive: it is there to be measured. In Quantum PCA, the quantum state ρ plays an active role in its own analysis. The quantum state ρ itself is used to implement the unitary operator $e^{i\rho t}$, a Hamiltonian that acts on copies of itself to reveal its own variance properties. This remarkable introspective capability, where the quantum particles effectively self-analyze, is unseen in classical systems.

5.2.3   POTENTIAL IMPACT OF QUANTUM PCA ON ROBOTICS

In robotics, PCA has been used for belief compression in POMDP planning [Roy and Gordon, 2003], simultaneous mapping and localization models [Brunskill and Roy, 2005], and robot environment modeling [Vlassis and Krose, 1999]. These are areas that may expect a speedup in computational processes. In addition, the self-analysis aspects of Quantum PCA may lead to some interesting development of sensor technology.

5.3   QUANTUM REGRESSION

Regression analysis aims to understand the relationships between statistical variables $\{(x_1, y_1), \ldots, (x_n, y_n)\}$. The goal is to approximate the statistical relation between the variables using a function f such that $\{(x_1, f(x_1)), \ldots, (x_n, f(x_n))\}$ resembles the original training data. With the approximation of f, values outside the training data such as $(z, f(z))$ can be extrapolated.

Regression methods are ubiquitous in robotics since so much of real-world physics can be modeled with continuous statistical processes. Large regression problems with many data points and variables may be solved exponentially faster using quantum methods, allowing robots to handle more complicated environments.

5.3.1   LEAST SQUARES FITTING

Least squares fitting is a common algorithm for classical regression. A function of the form:

$$f(x, \lambda) = \sum_{j=1}^{M} f_j(x)\lambda_j \tag{5.10}$$

is fit where $f(x, \lambda) : \mathbb{C}^{M+1} \mapsto \mathbb{C}$ and λ_i is a component of the weight vector λ. The fitting is done with the criterion of minimizing the magnitude of the least squares error:

$$\begin{aligned} E &= \sum_{i=1}^{N} |f(x_i, \lambda) - y_i|^2 \\ &= |F\lambda - y|^2. \end{aligned} \tag{5.11}$$

The fit parameters are found by applying the Moore-Penrose pseudoinverse of F (denoted F^+). The solution is given by:

$$\lambda = F^+ y = \left(F^\dagger F\right)^{-1} F^\dagger y. \tag{5.12}$$

5.3.2   QUANTUM APPROACHES TO CURVE FITTING

HHL (named after its co-inventors: Aram Harrow, Avinatan Hassidim, and Seth Lloyd) [Harrow et al., 2009] is a quantum algorithm for efficiently analyzing a linear system of the form:

$$Fx = b \tag{5.13}$$

where F is a sparse Hermitian matrix, x is a vector of predictor variables, and b is the $N \times 1$ vector of predicted variables. HLL computes the expectation value $< x|M|x >$ of an arbitrary poly-size Hermitian operator M in $O(\frac{s^4 \kappa^2 \log N}{\epsilon})$ steps where κ is the matrix F's condition number, s is the maximum number of nonzero elements in any row or column of F, and ϵ is the acceptable error distance to the optimal solution.

The key insight of Wiebe et al. [2012] is that HLL can be adapted to the problem of curve fitting, so that one can efficiently obtain the quality of the fit without having to learn the fit parameters. The HLL algorithm is used to construct the quantum state:

$$I(F)^{-1}I(F^{\dagger})|y\rangle \otimes |y\rangle \tag{5.14}$$

where I is an isometry superoperator which maps a matrix X to another matrix of the following form:

$$I : X \mapsto \begin{bmatrix} 0 & X \\ X^{\dagger} & 0 \end{bmatrix}. \tag{5.15}$$

The time complexity of implementing the quantum state in Equation 5.14 within error ϵ is $O(\frac{s^3\kappa^4 \log N}{\epsilon})$ with high probability. A swap test is then used to efficiently determine the accuracy of the fit. The swap test can distinguish $|y\rangle$ from $I(F)|\lambda\rangle$ and return "1" if the states are different. The overlap of the two quantum states (that determines the fit quality) can be statistically measured by repeating the swap test multiple times.

Limitations of Wiebe et al. [2012] are the assumptions of F being sparse and Hermitian and that the full fitted curve is not determined efficiently. Wang [2014] attempts to provide a quantum curve fitting algorithm that can work for the general case, though their running time is poly-logarithmic in the number of samples but not in the dimensionality of the data (a fundamental tradeoff between the two contrasting approaches). The approach in Wang [2014] leverages the density matrix exponentiation technique developed by Lloyd et al. [2014] that can simulate general non-sparse Hamiltonians. Wang [2014] proposes simulating the Hamiltonian $e^{iFF^T t}$ (where $i = \sqrt{-1}$) to create an algorithm that is theoretically exponentially faster than classical curve-fitting. It remains to be experimentally verified whether this complexity result is true in practice with real capability to simulate non-sparse Hamiltonians.

5.3.3 POTENTIAL IMPACT OF QUANTUM REGRESSION ON ROBOTICS

Given the prevalence of regression methods in robotics, there are numerous applications. Since so many computational modeling problems are continuous in robotics, a major benefit would be given to large regression problems with many data points and variables.

5.4 QUANTUM CLUSTERING

Clustering is another type of exploratory statistical analysis. In clustering, a set of objects is divided into several subgroups (clusters) such that each contains similar objects. Quantum clustering represents an exponential speedup on classical equivalents (in some cases), enabling the clustering of significantly larger datasets. For example, quantum cluster analysis may one day scale clustering to run over multi-exabyte data sets of perceptual data.

5.4.1 CLASSICAL CLUSTER ANALYSIS

The aim of clustering is to identify a set of K clusters (or partitions) $C = \{C_1, C_2, \ldots C_K\}$ which coherently divide a set of input data points $x = \{x_1, x_2, \ldots, x_n\}$ such that each data point is assigned to one or more clusters. A generic clustering algorithm is described by the procedure in Algorithm 5.1. A variety of clustering methods exist which employ different measures of similarity and distance used to assign data points to their "closest" cluster.

Algorithm 5.1 Generic Clustering Algorithm

Input : Data points $x = \{x_1, x_2, \ldots, x_n\}$, A set of initial K cluster centroids $Z = \{Z_1, Z_2, \ldots, Z_K\}$ (generally cluster guesses, often generated at random)
Output : K stable cluster centroids

1: **repeat**
2:   Form K clusters by assigning each point to the "closest" cluster Z_k
3:   Recompute the centroid of each cluster based on the new assignment of data points.
4: **until** Centroids do not change.
5: **return** K stable cluster centroids $Z = \{Z_1, Z_2, \ldots, Z_K\}$

One common method, k-means clustering, assigns data points to clusters by minimizing the Within-Cluster Sum of Squares (WCSS).

$$WCSS = \underset{C}{\operatorname{argmin}} \sum_{i=1}^{K} \sum_{x \in C_i} ||x - \mu_i||^2. \tag{5.16}$$

This method calculates the sum of squares within each cluster individually, by computing the squared distance from each data point x to its cluster's geometric mean μ_i. The individual data point squared distances are then summed together. The algorithm attempts to find an assignment of data points to clusters to minimize this cost.

Another method, k-medians clustering, optimizes data to cluster assignment using the geometric median instead. k-medians uses the cost function:

$$\underset{C}{\operatorname{argmin}} \sum_{i=1}^{K} \sum_{x \in C_i} |x - m_i|. \tag{5.17}$$

The median of cluster C_i is denoted as m_i. $|x - m_i|$ is the measure of Manhattan distance from each data point to the geometric median of the cluster to which it belongs. The cost function aims to find the minimum cost solution across all possible clusterings of the points.

5.4.2   QUANTUM CLUSTER ANALYSIS

Attempts have been made to extend classical clustering to quantum clustering. One method of quantum clustering utilizes Grover's algorithm in conjunction with a designed oracle function [Aïmeur et al., 2013]. Consider all pairs of points x_i and x_j within cluster C_k. While Aïmeur et al. [2013] did not specify the oracle's formula or quantum circuit implementation, their approach can generically be understood as applying the Hadamard gate on the superposition of all the pairs of indexes. The oracle would then output a superposition including all the triples $|i, j, Dist(x_i, x_j)\rangle$ as quantum states which allow for speeding up the calculation of distances between pairs of points using quantum subroutines.

For example, Grover's algorithm can be used to calculate the minimum value of a subset of distance values. If the subset is the list of centroids, this quantum subroutine can be used to compute Step 2 in Algorithm 5.1. If the subset is the list of data points assigned to a centroid, this quantum subroutine can be used to compute Step 3 in Algorithm 5.1. Overall, these quantum subroutines are hypothesized to result in a quadratic speedup for quantum clustering.

Another approach to quantum clustering utilizes adiabatic computation [Lloyd et al., 2013]. This formulation of a quantum k-means algorithm constructs the quantum state:

$$|\psi\rangle = \frac{1}{\sqrt{M}} \sum_{c, j \in c} |c\rangle |j\rangle \tag{5.18}$$

where c is a cluster, j represents a point in cluster c, and M is the number of data points. This represents the output of k-means clustering in quantum state form and can be measured to produce the solution. To construct this state, Lloyd et al. [2013] begins by initializing the basic quantum state:

$$\frac{1}{\sqrt{Mk}} \sum_{c'j} |c'\rangle |j\rangle \left(\frac{1}{\sqrt{k}} \sum_c |c\rangle |i_c\rangle \right)^{\otimes d} \tag{5.19}$$

where k is the number of clusters, M is the number of data points, and $|c'\rangle |j\rangle$ represents an assignment initialization guess. d copies of the seed state $\frac{1}{\sqrt{k}} \sum_c |c\rangle |i_c\rangle$ allow computing data point distances to cluster centers in superposition. Let $\vec{v}_j$ be a data point. Let $\vec{v}_{i_{c'}}$ be a cluster center. Then, using Hamiltonians, one can compute all $|\vec{v}_j - \vec{v}_{i_{c'}}|^2$ in superposition.

Two Hamiltonians are applied in succession. The first, H_1, evaluates distances and corresponds to Step 2 in Algorithm 5.1.

$$H_1 = \sum_{c'j} |\vec{v}_j - \vec{v}_{i_{c'}}|^2 |c'\rangle \langle c'| \otimes |j\rangle \langle j|. \tag{5.20}$$

The second, H_f, performs the cluster assignment. This corresponds to Step 3 in Algorithm 5.1.

$$H_f = \sum_{c'j} |\vec{v}_j - \vec{v}_{c'}|^2 |c'\rangle \langle c'| \otimes |j\rangle \langle j| \otimes I^{\otimes d}. \tag{5.21}$$

Both taken together construct the quantum state $|\psi\rangle$. The construction of the quantum state potentially allows for an exponential speedup over classical approaches.

5.4.3   POTENTIAL IMPACT OF QUANTUM CLUSTERING ON ROBOTICS

In robotics, classical clustering has been used to understand image data [Martínez and Vitria, 2001], trajectory data [Aleotti and Caselli, 2005], and robot experience [Oates et al., 2000]. These are all domains that will receive a speedup from quantum clustering. A robot using quantum clustering may be able to understand more complex percepts and patterns in noisy sensor data.

5.5   QUANTUM SUPPORT VECTOR MACHINES

The Support Vector Machine (SVM) is a powerful classification tool that takes in data with supervised labels and learns to classify previously unseen instances. In robotics, SVM has been used for learning robotic grasping models [Pelossof et al., 2004], place recognition models [Pronobis et al., 2008], data fusion methods [Xizhe et al., 2004], and object recognition [Ji et al., 2012].

5.5.1   CLASSICAL SVM ANALYSIS

The simplest Support Vector Machine model takes as input a set of training instances $\{(x_1, y_1), (x_2, y_2), \ldots, (x_N, y_N)\}$ where x_i are feature vectors and $y_i \in \{-1, +1\}$ are binary classes to which data points belong. An SVM attempts to fit a linear separating hyperplane to the data that maximally separates the data from the two classes.

Formally, a hyperplane in $\mathbb{R}^d$ is $w^T x - b = 0$ where w is the normal vector to the hyperplane and b is a bias parameter that helps determine offset from the origin. The margin equation for the two classes is given by $y_i(w^T x_i - b) \geq 1$ for $i = 1, \ldots, N$. The distance between the two classes is given by $||\frac{2}{w}||$, and the goal is to minimize w to get the maximum margin between the two classes. This optimization problem is represented by $\underset{w,b}{\operatorname{argmin}} \frac{1}{2}\|w\|^2$.

Since most classification problems are not precisely linearly separable, soft margin SVMs are commonly used to find a solution. A soft margin relaxes the assumption that positive and negative training examples need to be perfectly linearly separable by a hard margin. In the soft margin SVM, the introduction of nonnegative slack variables ϵ_i into the objective function that measure the degree to which a data point deviates from the margin allows element densities to overlap. The soft margin SVM objective is given by $\min \frac{1}{2}\|w\|^2 + C \sum_i \epsilon_i$.

Kernel functions can be used to automatically map data from an input space to a higher dimensional space. The key idea is to transform the data using a function $\phi(x)$ that retains many properties of the inner product but is nonlinear. The Kernel function nonlinearly maps the input data to a higher dimensional space where a linear separator can be fit with greater ease.

Given all these constraints, the following optimization problem is obtained (using the Lagrangian of the dual problem):

$$\max_{\alpha_i} \sum_{i=1}^{N} \alpha_i - \frac{1}{2} \sum_{i,j} \alpha_i \alpha_j \, y_i \, y_j \, K(x_i, x_j)$$

$$\text{subject to} \quad 0 \leq \alpha_i \leq C \quad \sum_{i=1}^{N} \alpha_i \, y_i = 0$$

(5.22)

where $K(x_i, x_j) = \phi(x_i)^T \phi(x_j)$.

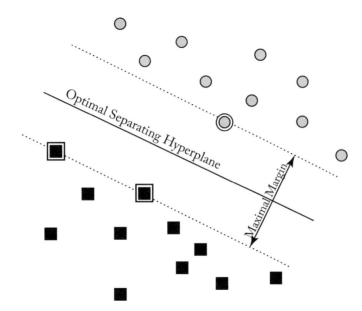

Figure 5.1: Example support vector machine and fit separating max-margin hyperplane on two class data.

The α_i are learned weights for points. The few α_i that are non-zero correspond to the "support vectors" that define the margin. The C constraint is a form of regularization that follows from the derivation of the slack variables, trading off between maximizing the margin and allowing more softness on the margin. The Kernel function $K(x_i, x_j)$ allows the SVM to first transform the data to a nonlinear space before fitting the hyperplane.

For an arbitrary new data point x,

$$f(x) = \text{sign} \left(\sum_{i=1}^{N} \alpha_i \, y_i \, K(x_i, x) + b \right)$$

(5.23)

assigns the new point to its category (i.e., either the $+$ class or the $-$ class, based on the sign of the expression).

The SVM is classically useful on small to medium-size data sets but challenging to scale to larger data sets for real-time robotics. In addition, convex loss functions must often be used with the SVM (such as the well-known hinge loss) for the optimization to be tractable. However, convex functions can be sensitive to noise, especially class label (i.e., $+/-$) outliers [Anguita et al., 2003].

5.5.2   QUANTUM SVM ANALYSIS

Several quantum variants of the SVM have been proposed. In theory, a quantum SVM could scale to much larger data sets than the classical SVM. The quantum SVM would also not have to make convex approximations, potentially finding more robust decision boundaries that increase classification accuracy.

Anguita et al. [2003] proposes a Grover-style algorithm to perform an exhaustive search in the parameter and cost space for any arbitrary SVM objective function. Even non-convex SVM objective functions (such as the famous "0/1" loss function) can be used in the optimization. While generally this type of discrete optimization would be intractable classically, running Monte Carlo Random Search on a quantum computer could be useful with speedups from quantum parallelism.

Assume that the minimization problem has M different solutions. In a search space with N possible configurations, the probability of success of a Monte Carlo search after r iterations is: $P_m^{(r)} = 1 - (1 - M/N)^r$. The quantum approach exhibits an advantage whenever $22.5M < \sqrt{N} \ln 2$ [Anguita et al., 2003].

Neven et al. [2008] proposes an adiabatic optimization approach to the SVM where the SVM objective function is cast as a Quadratic Unconstrained Binary Program (QUBO). QUBO structures are well-optimizable by adiabatic optimization. The paper reports better generalization error may be possible for many problems.

Rebentrost et al. [2014] illustrates that an exponential speedup is possible when using the least squares formulation of the SVM. By using Hamiltonian simulation on an adiabatic processor as well as expressing the problem in the eigenbasis obtained by Quantum PCA [Lloyd et al., 2014], a quantum variant of the least squares SVM can be fit efficiently:

$$
\begin{aligned}
F &= \begin{pmatrix} 0 & 1^T \\ 1 & K + \gamma^{-1}I \end{pmatrix} = J + K_\gamma \\
J &= \begin{pmatrix} 0 & 1^T \\ 1 & 0 \end{pmatrix} \\
K_\gamma &= \begin{pmatrix} 0 & 0 \\ 0 & K + \gamma^{-1}I \end{pmatrix}
\end{aligned}
\tag{5.24}
$$

where K is the kernel matrix, γ is a regularization parameter, I is the identity matrix, and 1 is a vector of ones. By simulating the matrix exponential of F, an exponential speedup in SVM training is forecasted by this model [Rebentrost et al., 2014].

5.5.3   POTENTIAL IMPACT OF QUANTUM SVMS ON ROBOTICS

In robotics, SVM is used for problems such as learning robotic grasping models [Pelossof et al., 2004], place recognition models [Pronobis et al., 2008], data fusion methods [Xizhe et al., 2004], and object recognition [Ji et al., 2012]. One may expect to be able to build a classifier on a larger amount of data with a quantum SVM. Thus one may expect to be able to grasp objects that are more complex and navigate environments with more varied structure.

5.6   QUANTUM BAYESIAN NETWORKS

A Bayesian Network is a compact graphical representation of a complicated joint probability distribution function $P(X_1, \ldots, X_n)$. Often times, real-world probability distributions have a conditional independence structure that allows the factorization of the distribution into a few product terms. Bayesian Networks are one method of concisely representing such a factorization.

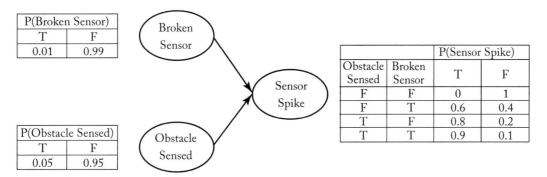

Figure 5.2: Example Bayesian network for diagnosing the health of a sonar range sensor.

Figure 5.2 shows an example Bayesian Network. The nodes represent variables, and the links between variables express conditional independence relations. A node in a Bayesian Network is conditionally independent of other nodes given its parent nodes. A Bayesian Network thus factorizes according to the following rule:

$$P(X_1, \ldots, X_n) = \prod_{i=1}^{n} P(X_i \mid \text{parents}(X_i)). \tag{5.25}$$

In the example shown in Figure 5.2, imagine a robot with a sonar range sensor. The sonar sensor allows the robot to detect obstacles in the environment, returning a spike in sensor signal if an obstacle is detected. However, the sonar sensor may also break (e.g., due to the robot falling on itself) and, if it is broken, the sonar sensor may spike spuriously. If a sensor spike is received, one may want to try to infer the likely cause. This situation can be modeled using a Bayesian Network.

Let B represent the event "Broken Sensor," O represent the event "Obstacle Sensed," and S represent the event "Sensor Spike." In our example, the network shows that either having a broken sensor and/or sensing an obstacle in the environment may lead to a spike in the sensor's signal. These factors can be used to predict the probability of getting a sensor spike. If one is obtained empirically, one can try to infer the likely causes.

The factorization of the probability distribution function shown in the Bayes Net is: $P(B, O, S) = P(S|B, O)P(B)P(O)$. The conditional probability tables (CPTs) express the parameters of the probability distribution. Notice that the factorization makes the probability distribution more succinct, requiring less memory to store the joint distribution.

A Bayesian Network, with defined structure and CPT parameters, can be used for inferring unobserved variables. Any joint probability in the overall probability distribution function can be calculated via the factorization and CPT parameters (for algorithms regarding this capability, see Koller and Friedman [2009]).

In robotics, Bayesian Networks have been applied to particle filtering [Doucet et al., 2000], structure learning in robot grasping [Song et al., 2011], 3D reconstruction [Delage et al., 2006], and robot localization [Zhou and Sakane, 2007], amongst other problems.

5.6.1   CLASSICAL BAYESIAN NETWORK STRUCTURE LEARNING

The problem of Bayesian Network Structure Learning (BNSL) is to learn both the network conditional independence structure and CPT parameters from data. Classically, there are many ways to tackle this problem. Methods include using independence tests between pairs of variables to establish variable relations, scoring the entire graph structure (using maximum likelihood estimation) and searching over structures, using Bayesian scores and a prior to bias the search to particular structures, and Bayesian model averaging with top models to fuse multiple models to find the optimal fit. For an in-depth treatment of classical BNSL, see Koller and Friedman [2009].

Classically, BNSL is an exponentially hard computational problem. While this does not change in the quantum world, one does expect a significant speedup in solving instances of BNSL and potential capability to learn larger Bayesian Network structures from data.

5.6.2   BAYESIAN NETWORK STRUCTURE LEARNING USING ADIABATIC OPTIMIZATION

O'Gorman et al. [2015] propose an algorithm for using quantum adiabatic optimization to perform BNSL. The general strategy is as follows:

Step 1 of the algorithm involves creating a bit string representation of a candidate graph structure. Define: $\mathbf{d} = (d_{ij}) \mid 1 \leq i < j \leq n, i \neq j$ as the string of bits of possible edges where

Algorithm 5.2 Quantum Adiabatic Optimization for Bayesian Network Structure Learning

Input : n variables
Output : QUBO instance for BNSL problem (solvable via adiabatic optimization)

1: Encode all digraphs using a set of Boolean variables, each of which indicates presence or absence of a directed edge.
2: Define a quadratic pseudo-Boolean function on those variables that scores encoded digraphs.
3: Encode additional resource and graphical constraints for optimization.
4: **return** Quadratic Unconstrained Binary Optimization (QUBO) instance that is defined over $O(n^2)$ variables when mapped from a BNSL instance with n variables.

$d_{ij} = 1$ indicates an edge from vertex X_i to vertex X_j, $d_{ij} = 0$ indicates no edge, and $d_{ii} = 0$. $G(\mathbf{d})$ specifies an encoded candidate directed graph.

Step 2 of the algorithm involves defining a quadratic pseudo-Boolean Hamiltonian function $H_{\text{score}}(\mathbf{d})$ such that $H_{\text{score}}(\mathbf{d}) = s(G(\mathbf{d}))$, the likelihood score for the proposed graphical structure given the data. A pseudo-Boolean function is a function $f : \mathbf{B}^b \to \mathbb{R}$ where $\mathbf{B} = \{0, 1\}$ is a Boolean domain and b is a nonnegative integer called the arity of the function.

Any pseudo-Boolean function has a unique multinomial form, so one can write:

$$H_{\text{score}}^{(i)}(d_i) = \sum_{j \subset \{1,\dots,n\}\setminus\{i\}, |J| \leq m} \left(w_i(J) \prod_{j \in J} d_{ji} \right)$$

$$w_i(J) = \sum_{l=0}^{|J|} (-1)^{|J|-l} \sum_{K \subset J, |K|=l} s_i(K).$$

(5.26)

Additional considerations of the optimization include incorporating a resource constraint on the maximum connectivity and a penalty on cycles in the graphical structure. The resource constraint is implemented since quantum adiabatic hardware is currently very limited in total qubit connections. For this constraint, the function $H_{\text{max}}^{(i)}$ is defined whose value is zero if variable X_i has at most m parents and positive otherwise. Cycles in the graph are also penalized using a $H_{\text{cycle}}^{(i)}$ term. This constraint leverages a theoretical computer science result that, if a tournament has any cycle, it has a cycle of length three. Thus, to penalize cycles, it is sufficient to penalize directed 3-cycles. The overall Hamiltonian is given by:

$$H = H_{\text{score}} + H_{\text{max}} + H_{\text{cycle}}.$$

(5.27)

The Hamiltonian ensures that the ground state of the system encodes the highest likelihood scoring directed acyclic graph with a maximum parent set size that can represent the Bayesian Network.

The algorithm described in O'Gorman et al. [2015] has been implemented on a D-Wave Two chip system installed at NASA Ames Research Center that allows problems up to seven Bayesian Network variables. Note that the implementation is not yet competitive, as classical methods are able to deal with many tens of Bayesian Network variables. A polynomial speedup on BNSL is expected with quantum adiabatic optimization, though is yet to be demonstrated with current devices.

5.6.3 POTENTIAL IMPACT OF QUANTUM BAYESIAN NETWORKS ON ROBOTICS

Many classical robotic representations are Bayesian. Problems such as robotic localization and mapping have been cast within a Bayesian framework with much practical success. Bayesian Networks have been applied to particle filtering [Doucet et al., 2000], structure learning in robot grasping [Song et al., 2011], 3D reconstruction [Delage et al., 2006], and robot localization [Zhou and Sakane, 2007], amongst other problems. Current Bayesian Network methods are limited by the capability to learn complex graphical structures. If Quantum Bayesian Networks are able to scale to larger graphical structures, a richer representation of robotic interactions with an environment could be captured.

5.7 QUANTUM ARTIFICIAL NEURAL NETWORKS

Artificial Neural networks (ANNs) are a model of computation loosely inspired by the structure of the brain. ANNs consist of connected neurons that, with enough layers, can serve as a powerful non-linear classifier or feature extractor. In robotics, ANNs have helped mobile robot perception [Pomerleau, 1993] and control of robot manipulators [Lewis et al., 1998], amongst a growing number of applications.

5.7.1 CLASSICAL ARTIFICIAL NEURAL NETWORKS

An artificial neural network is a network of units ("neurons") connected by weighted edges ("synapses"). The network receives an input, applies a complex transformation to the input based on its connections, and produces a response called the "activation."

The simplest case, called a "Perceptron," consists of only a single neuron [Rosenblatt, 1958]. A single neuron works as a binary classifier. The activation function models whether the neuron fires or not:

$$f(x) = \begin{cases} 1 & \text{if } w^T x + b > 0 \\ 0 & \text{otherwise} \end{cases} \tag{5.28}$$

where x is the input vector in $\mathbb{R}^d$, w is a weight vector, and b is the bias term. Training of the classical perceptron is facilitated via the "delta rule," a form of gradient descent where each weight w_j is updated in proportion to the error it produces in prediction:

$$\Delta w_j = \gamma(y_i - f(x_i))x_j \qquad (5.29)$$

where y_i is the true label of the data point and γ is a predefined learning rate that can be varied throughout training.

The perceptron model suffers from the "XOR" problem [Minsky and Papert, 1969]. Training will not converge if the training examples are not linearly separable with respect to their classes. The solution is to use multiple layers in the neural network.

Feedforward networks, perhaps the most common type of multi-layer artificial neural network, assume connections only go in one direction with no directed cycles. Typical feedforward networks have an input layer, some number of neurons in a hidden layer, and an output layer for classification. Feedforward networks can be trained using the famous backpropagation algorithm which updates weights in the network in proportion to the error they produce [Rumelhart et al., 1988].

A Hopfield Network [Hopfield, 1982], another type of multi-layer artificial neural network, is an "associative memory" which recalls a canonical pattern previously generalized from training examples. Given an input pattern, it automatically generalizes to the output without requiring explicit addressing of the training examples.

Structurally, a Hopfield Network is an assembly of fully connected perceptrons with no self-loops. K perceptrons have $K(K - 1)$ interconnections. The perceptrons send signals around to each other until convergence in either a synchronous or asynchronous function. In the simplest model, the weights are assumed to be symmetric with the state of a single neuron s_i being either $+1$ or -1 determined by:

$$s_i = \begin{cases} +1 & \text{if } \sum_j w_{ij}s_j \geq \theta_i \\ -1 & \text{otherwise} \end{cases} \qquad (5.30)$$

where θ_i is a predefined neuron firing threshold. The "Energy" of the Hopfield Network is determined as:

$$E = -\frac{1}{2}\sum_{i,j} w_{ij}s_is_j + \sum_i \theta_i s_i. \qquad (5.31)$$

Interestingly, updating weights in the network to learn a pattern always means the energy quantity remains the same or decreases. This dissipative nature of neural computation makes implementation of quantum artificial neural networks challenging.

Boltzmann Machines [Hinton and Sejnowski, 1986] can be viewed as a stochastic variant of the Hopfield network. A Boltzmann Machine consists of two sets of units: visible units (v) which encode the input/output of the network and hidden units (h) which capture complex correlations between input and output. Like Hopfield units, these units can be "on" (1) or "off" (0). The probability density function of a Boltzmann Machine for a given configuration of visible and hidden nodes is given by:

$$P(v,h) = \frac{e^{-E(v,h)}}{Z}$$

$$E(v,h) = -\sum_i v_i b_i - \sum_j h_j d_j - \sum_i \sum_j w_{ij}^{vh} v_i h_j - \sum_i \sum_j w_{ij}^v v_i v_j - \sum_i \sum_j w_{ij}^h h_i h_j$$

$$(5.32)$$

where $E(v,h)$ is the network's energy function, b and d are bias vectors that provide a penalty for a unit taking a value of 1, and w^{vh}, w^v, and w^h are tunable weight vectors. Z is a normalization factor, also known as a partition function.

Restricted Boltzmann Machines (RBMs) [Smolensky, 1986] are a class of Boltzmann machines where visible units are not directly connected to other visible units, and hidden units are not directly connected to other hidden units. Thus, the visible units and hidden units form a bipartite graph and $w^h = w^v = 0$.

Long training time served as a major hurdle for multi-layer artificial neural networks. Deep Learning [Bengio, 2009, Bengio et al., 2007, Hinton et al., 2006, LeCun et al., 2010] leverages computational tricks such as learning by layers, feed-forward approximations, parallel architectures, and GPU acceleration to speed up computation. For example, Boltzmann machines have traditionally been hard to train since computing the partition function is $\#P$-hard. Deep Restricted Boltzmann Machines (dRBMs) [Le Roux and Bengio, 2008, Salakhutdinov and Hinton, 2009] require that each layer of hidden units is connected only to units in adjacent layers. This allows for the use of a contrastive divergence approximation to efficiently train the network in a greedy layer-by-layer fashion. The classical approximation can lose accuracy, which may be improvable by quantum approaches.

5.7.2   QUANTUM APPROACHES TO ARTIFICIAL NEURAL NETWORKS

Quantum variants of the Hopfield Network (called Quantum Associative Memories), Perceptrons, and Feedforward networks have been suggested. Quantum Artificial Neural Networks have several potential advantages to classical artificial neural networks. First, quantum artificial neural networks can likely have a more powerful representation capability by storing network weights in superposition. Second, speedups in optimization suggest potentially faster training times for backpropagation. Finally, quantum artificial neural networks may more easily find the optimum of a complicated non-convex objective function than classical approaches.

Quantum Associative Memories

Quantum Associative Memories [Ventura and Martinez, 2000] are the quantum extension to Hopfield Networks. Quantum Associative Memories store patterns in superposition. While classical Hopfield Networks provide $O(d)$ memory storage using d neurons, Quantum Associative Memories provide $O(2^d)$ storage using d neurons. Quantum Associative Memories store their

N patterns in an equally-weighted superposition:

$$|M\rangle = \frac{1}{\sqrt{N}} \sum_{i=1}^{N} |x_i\rangle. \tag{5.33}$$

Since measurement can only return one value, the memory $|M\rangle$ has to be periodically cloned [Duan and Guo, 1998]. To avoid contradicting the non-cloning theorem, the cloning of memory is approximate and thus imperfect.

Assuming the memory can be maintained, there are two key approaches to pattern matching and classification of an unseen data point. The first is a modified Grover's algorithm [Ventura and Martinez, 2000], augmented with extra rotations to allow efficient search for patterns over the database when not all patterns are represented. The second (known as "Probabilistic Quantum Memories" [Trugenberger, 2001]) is to calculate the Hamming distance between the input pattern and the instances in memory. The following Hamiltonian measures the number of bits that the input pattern and a stored memory instance differ:

$$H = d_m \otimes (\sigma_z)_c$$
$$d_m = \sum_{k=1}^{n} \left(\frac{\sigma_z + 1}{2} \right)_{m_k} \tag{5.34}$$

where σ_z is the Pauli Z matrix, m_k is a memory instance, n is the length of a memory instance numeric vector, and c is a control bit. By applying the Hamiltonian to the input pattern and memory instances, the probability of obtaining a pattern x_k for input i is:

$$P(x_k) = \frac{1}{N * P(|c\rangle = |0\rangle)} \cos^2 \left(\frac{\pi}{2L} d(i, x_k) \right) \tag{5.35}$$

which peaks around the patterns with the smallest Hamming distance ($|c\rangle$ is a control bit). This strategy helps identify the pattern closest to the input.

The energy function of the Hopfield network [Equation (5.31)] resembles the Hamiltonian of a spin glass or Ising model in the quantum world. This makes adiabatic quantum computing a prime target paradigm for implementation of quantum artificial neural networks. Adiabatic quantum computing has also been used to implement quantum associative memories. Neigovzen et al. [2009] demonstrate a quantum associative memory with two qubits in a liquid-state nuclear magnetic resonance system. This approach is further studied by Seddiqi and Humble [2014].

Quantum Perceptron and Feedforward Networks

A key challenge with building a quantum perceptron is incorporating nonlinearity. Activation functions in the classical perceptron create nonlinearities in the mathematical model. Individual quantum states, however, can only evolve according to linear, unitary, and reversible dynamics. The literature suggests two key approaches for incorporating nonlinearity into quantum perceptrons.

Lewenstein [1994] and Zak and Williams [1998] suggest using quantum measurement to play the role of the threshold function in the quantum perceptron. In quantum mechanics, a measurement collapses a superposition to one state from a distribution of possible states. This serves as an effective thresholding mechanism:

$$(a_1, \ldots, a_N) \rightarrow (0, \ldots, 1, \ldots, 0) \text{ with probability } |a_i|^2. \tag{5.36}$$

From this mechanism, one can define an update rule on quantum states that is similar to the classical neuron firing rule:

$$a_{t+1} = \sigma(Ua_t) \tag{5.37}$$

where U is a unitary operator, a_t is a previous state, and the function $\sigma(x)$ performs a measurement and reinitializes the quantum state a_{t+1}. Narayanan and Menneer [2000] suggests that the principle of alternating classical and quantum components can be used not just in a perceptron but can be scaled to a multilayer network of neurons.

Another approach to incorporating nonlinearity into quantum mechanics is via the Feynman Path Integral [Faber and Giraldi, 2002]:

$$|\psi(x_f, T)\rangle = \int_{(x_0, 0)}^{(x_f, T)} D[x(t)] e^{\frac{i}{\hbar} \int_0^T d\tau [\frac{1}{2} m\dot{x} - V(x)]} |\psi(x_0, 0)\rangle \tag{5.38}$$

where $|\psi(x_0, 0)\rangle$ is the initial state of the quantum system at time $t = 0$, $|\psi(x_f, T)\rangle$ is the system at time $t = T$, $\hbar$ is Plank's constant, m is the mass, and V is the potential energy. Nonlinearity can be adjusted through model training by using the $V(x)$ function as well as via the coupling of the quantum system to its environment. The environment surrounding the quantum system is a potential design choice that can introduce nonlinearity. Behrman et al. [1996] demonstrate a possible implementation strategy for this approach via solid-state quantum dot hardware, which is discussed further in Section 7.4.

Quantum Deep Learning

Training Boltzmann Machines (BMs) generally requires approximation since calculating the gradients of the objective function (which involves evaluation of the notorious partition function Z) is intractable. Practical applications of Deep Restricted Boltzmann Machines (dRBMs), for example, often utilize mean field [Bengio, 2009] or contrastive divergence [Hinton, 2002] approximations to the gradients which allow some sizes of dRBMs to be trained on classical computers.

Wiebe et al. [2014] explores whether quantum strategies can be used to enable efficient training of BMs and dRBMs. They provide some alternative approaches to the approximations previously posited for these networks by employing quantum sampling. By optimizing their procedure with quantum amplitude estimation (a variant of Grover's algorithm), they show BMs can be trained for achieving a fixed sampling error in modeling with quadratically fewer operations than classical algorithms. In addition, their quantum approach shows improved accuracy (in simulation) for learning higher quality models, especially for large BMs on noisy data.

Interestingly, the quantum deep learning approach circumvents the need for a dRBM topology of the network. While classically, the dRBM is perhaps one of the most scalable of BM-style models, it is also a fairly restrictive model of data. In the quantum world, approaches may scale to training true BMs with fewer required assumptions. For instance, the contrastive divergence approximation for gradients may not be necessary with quantum annealing.

Attempts have been made to evaluate the feasibility of RBMs that use quantum sampling on the D-Wave hardware [Adachi and Henderson, 2015, Denil and De Freitas, 2011, Dumoulin et al., 2013]. Implementation is challenging since performance can be constrained by the limited connectivity of qubits in the machine and faulty qubits from imperfections introduced during the manufacturing process [Dumoulin et al., 2013]. Preliminary results from these studies suggest that quantum sampling for Deep Neural Networks can obtain similar or better accuracy with fewer iterations than the classical contrastive divergence / dRBM training procedure on data sets like MNIST [Adachi and Henderson, 2015]. More tests are needed to determine whether the results are due to quantum effects, whether the quantum annealing approach can improve results on other data sets, and to determine how the approach scales relative to the classical training procedure.

Further Reading

The literature surrounding quantum artificial neural network and its theoretical building blocks is vast, and there are multiple excellent literature summaries of quantum artificial neural networks. For further reading on the topic, view Benedetti et al. [2015], Ezhov and Ventura [2000], Faber and Giraldi [2002], Schuld et al. [2014], Wittek [2014].

5.7.3 POTENTIAL IMPACT OF QUANTUM ARTIFICIAL NEURAL NETWORKS TO ROBOTICS

Artificial neural network (especially Deep Learning) approaches are currently popular in robotics for problem domains including computer vision [LeCun et al., 2010] and robotic reinforcement learning [Mnih et al., 2015]. Quantum Artificial Neural Networks may provide speedup to deal with more complex representations in these and other robotic problems.

5.8 MANIFOLD LEARNING AND QUANTUM SPEEDUPS

High dimensional data can be extremely difficult to interpret. Manifold learning leverages the idea that oftentimes data set dimensionality is only artificially high. Data may lie along a low-dimensional manifold embedded in a high-dimensional space.

An example of a manifold is the famous swiss roll (shown in Figure 5.3). The swiss roll data is 3D but only artificially so, as it is a 2D plane that has been wrapped into a spiral. The data is actually a 2D manifold in a 3D space. The goal of manifold learning is to find the lower-dimensional representation of the higher-dimensional data.

Swiss Roll Manifold

Figure 5.3: Example "Swiss Roll" manifold: A 2D plane embedded in a 3D space.

5.8.1   CLASSICAL MANIFOLD LEARNING

Isometric Feature Mapping (Isomap) [Tenenbaum et al., 2000] is a popular algorithm for manifold learning. The two steps of the algorithm are shown in Algorithm 5.3.

Algorithm 5.3 Isomap Algorithm

Input : Data points as a $N \times d_0$ data matrix, k (integer number of nearest neighbors to use in estimating geodesic distances)
Output : Lower-dimensional representation of N input points

1: Estimate the geodesic distances (distances along a manifold) between the data points in the input using shortest-path distances on the data set's k-nearest neighbor graph.
2: Use Multi-Dimensional Scaling (MDS) to find points in a low-dimensional Euclidean space whose inter-point distances match the distances found in step 1.
3: **return** Lower-dimensional representation of N data points as a $N \times d$ data matrix where the new dimensionality $d \leq d_0$.

Isomap makes the local linearity assumption that, if the manifold is smooth enough, the Euclidean distance between nearby points in the high-dimensional space is a good approximation to distances along the manifold. The Isomap algorithm first constructs a k-nearest neighbor graph that is weighted by Euclidean distances. A shortest path algorithm (such as Dijkstra's or Floyd's) is used to output estimates for all other inter-point distances. The result is a dissimilarity matrix $D \in \mathbb{R}^{n \times n}$ of distances between all points.

Afterward, Multidimensional Scaling (MDS) is used to construct a set of points in a lower-dimensional space whose inter-point distances match those in the higher-dimensional space. MDS solves the problem: given a matrix $D \in \mathbb{R}^{n \times n}$ of dissimilarities, construct a set of points whose inter-point Euclidean distances match those in D closely. The output is a lower-dimensional representation of the higher-dimensional data.

Other forms of manifold learning include Local Linear Embedding [Saul and Roweis, 2003] which attempts to capture local geometric features of points, Laplacian Eigenmaps [Belkin and Niyogi, 2001] which leverages the graph Laplacian to reveal information about the data, and Semidefinite Embedding [Weinberger et al., 2004] which uses semidefinite programming to approximate Gram matrices to convert between high- and low-dimensional spaces. For a more in-depth treatment of manifold learning techniques, the reader is encouraged to browse [Cayton, 2005].

Classical manifold learning has scalability challenges. Building and analyzing the neighborhood graph and calculating inter-point distances can be computationally demanding. Many manifold learning techniques can only handle hundreds to low thousands of data points without parallelization. Much work is done to optimize these algorithms to run on larger data sets. Furthermore, many manifold learning methods assume convex parameter spaces, though this is not always true of manifolds in the real world.

5.8.2   QUANTUM SPEEDUPS FOR MANIFOLD LEARNING

Quantum methods may help scale manifold learning to larger data sets. Dürr et al. [2004] develops a quantum subroutine quant_find_smallest_values which, with high probability using a variant of Grover's algorithm, can find the k closest neighbors of a point in $O(\sqrt{kn})$ time if there are n points in the database. This hypothetically allows the calculation of inter-point distances faster on a quantum computer than classically possible.

With high probability, the algorithm in Aïmeur et al. [2013] constructs the neighborhood graph of a set of n points in $O(\sqrt{k}n^{\frac{3}{2}})$ time. Classically, if one uses an arbitrary metric and if the only information available is the distance between pairs of points, $\Omega(n^2)$ time is required. With access to all the D features in the point, a KD-tree approach can provide a speedup to $O((k + D)n \log n)$. However, the fastest classical result is not as theoretically fast as the quantum result.

5.8.3   POTENTIAL IMPACT OF QUANTUM MANIFOLD LEARNING ON ROBOTICS

The theory of how classical manifold learning affects robotics is still in its infancy. Thus it may be premature to think about the application of quantum manifold learning to robotics. However, classical manifold learning has proven useful for problems in robot navigation [Keeratipranon et al., 2006], dynamic visual tracking [Qiao et al., 2010], motion synthesis [Havoutis and Ramamoorthy, 2010], and is being tried on an increasing number of problems in robotics. These are all important robotic problem domains that will get an important speedup from quantum approaches.

5.9   QUANTUM BOOSTING

The idea behind boosting is to create a "strong" classifier out of many simpler "weak" classifiers. Boosting is motivated by the observation that simple classifiers can do unexpectedly well for many tasks. Having a diversity of simple classifiers and fusing them in a way that explicitly seeks out models that complement each other in reducing classification error can be a powerful strategy. In robotics, boosting has been applied to structured prediction for imitation learning [Ratliff et al., 2007], improving accuracy for semantic place classification [Rottmann et al., 2005], and learning loop closures [Walls and Wolcott, 2011].

5.9.1   CLASSICAL BOOSTING ANALYSIS

A popular boosting algorithm, Adaboost [Freund et al., 1999], works by trying to iteratively improve classification accuracy on previously incorrectly classified training instances (Algorithm 5.4). Adaboost can be a powerful algorithm to improve classification accuracy, though care must be taken to avoid overfitting to the training data. A key result is that the error rate (on training data) of the combined "strong" classifier approaches zero exponentially fast in the number of iterations if the weak classifiers do at least significantly better than random guessing [Freund et al., 1999].

Adaboost attempts to find a function:

$$f(x) = \sum_{s=1}^{S} w_s h_s(x) \tag{5.39}$$

where $f(x)$ weights the individual weak classifier predictions $h_s(x)$ to come up with the optimal predictor. To do so tractably, Adaboost minimizes a convex loss function over a convex set of functions. The loss Adaboost minimizes is:

$$L(y_i, f(x_i)) = e^{-y_i f(x_i)}. \tag{5.40}$$

The convexity assumptions can limit the modeling power of classical Adaboost. In addition, a key challenge with applying boosting style approaches in robotics has been getting ensemble methods

Algorithm 5.4 Adaboost Algorithm

Input : N training instances $\{(x_1, y_1), \ldots, (x_N, y_N)\}$
Output : "Strong" classifier from many "weak" classifiers

1: Initially assign a weight $w_i = \frac{1}{N}$ to each of N training instances.
2: At each iteration, a subset of training samples is drawn (with replacement). The selection probability of a training example is equal to its weight.
3: Train classifier ensemble on sample. Each classifier h_t (at iteration t) makes a prediction for each data point, and the measurement error of the classifier is computed. The error of one of the classifiers at iteration t is:
$$E_t = \sum_{i=1}^{N} w_i 1(y_i \neq h_t(x_i))$$
where $1(z)$ is the indicator function.
4: Adjust weights of instances so that incorrectly classified examples have higher weight and correctly classified examples have lower weight. A weighting rule often used is: set $w_{i,t} = w_{i,t-1} \times \frac{E_t}{1-E_t}$ for correctly classified instances, while keeping the weight of incorrectly classified instances the same. All weights are then re-normalized to form a probability distribution. The classifiers are also assigned weights: $\alpha_t = \frac{1}{2} \log \frac{1-E_t}{E_t}$, which upweights the output of simple classifiers in the ensemble that have low error rate when making predictions.
5: Repeat steps 2-4 for some number T of iterations or some error convergence criterion.
6: **return** Weighted combined classifier that can be used to make predictions on new data points.

to work well in real time. As more classifiers are used in an ensemble, the potential modeling power of Adaboost is improved for complicated world percepts. However, for a robotic system, it may be computationally intensive to evaluate a large number of classifiers on embedded hardware in real time. Quantum approaches to boosting may help alleviate some of these challenges.

5.9.2   QBOOST

Quantum Boosting (QBoost) [Neven et al., 2008] allows both the regularization term and loss function to be non-convex. By using adiabatic optimization and discrete search over the parameter space, the convexity assumption can be avoided. Note that the loss function needs to be quadratic (since adiabatic optimization is based on QUBOs) but can be non-convex.

Given a set of weak learners $\{h_s | h_s : \mathbb{R} \to \{-1, 1\}, s = 1, 2, \ldots, S\}$, boosting can be written as:

$$\underset{w}{\mathrm{argmin}} \left(\frac{1}{N} \sum_{i=1}^{N} \left(\sum_{s=1}^{S} w_s h_s(x_i) - y_i \right)^2 + \lambda ||w||_0 \right) \tag{5.41}$$

where $\lambda ||w||_0$ is a regularization term. Equation (5.41) can be expanded as:

$$\underset{w}{\mathrm{argmin}} \left(\frac{1}{N} \sum_{s=1}^{S} \sum_{z=1}^{S} w_s w_z \left(\sum_{i=1}^{N} h_s(x_i) h_z(x_i) \right) \right.$$
$$\left. - \frac{2}{N} \sum_{s=1}^{S} w_s \sum_{i=1}^{N} h_s(x_i) y_i + \lambda ||w||_0 \right). \tag{5.42}$$

Term 2 in Equation (5.42) upweights weak learners correlated with the correct output label. Term 1 in Equation (5.42) downweights weak learners correlated with each other. The resulting QUBO equation can be discretized and solved using quantum annealing.

5.9.3 POTENTIAL IMPACT OF QUANTUM BOOSTING ON ROBOTICS

In robotics, boosting has been applied to structured prediction for imitation learning [Ratliff et al., 2007], improving accuracy for semantic place classification [Rottmann et al., 2005], and learning loop closures [Walls and Wolcott, 2011]. These are all domains that may benefit from a quantum boosting algorithm.

5.10 CHAPTER SUMMARY

<u>**Chapter Key Points**</u>

- Quantum approaches to machine learning utilize a variety of quantum operating principles such as quantum inner products, Hamiltonian simulation, adiabatic optimization, and Grover's search to obtain benefits (in theory) over classical machine learning approaches.

- Many machine learning methods obtain speedups. Some methods (such as Quantum PCA) obtain new capabilities. Others (such as quantum SVM or quantum deep learning) may become more general in their modeling power.

- Developed quantum approaches, if and when they work, may be useful for a broad array of robotic problem domains and applications.

Table 5.1: Summary of quantum operating principles in quantum machine learning

Algorithm	Quantum Operating Principles (QOPs) Used	Expected Speedup (from classical)
Principal Component Analysis	Hamiltonian Simulation [Lloyd et al., 2014]	Exponential
Regression/Curve Fitting	Quantum Swap Test [Wiebe et al., 2012] Hamiltonian Simulation [Wang, 2014]	Exponential
Clustering	Grover's Algorithm [Aïmeur et al., 2013] Adiabatic Theorem [Lloyd et al., 2013]	Quadratic [Aïmeur et al., 2013] Exponential [Lloyd et al., 2013]
Support Vector Machines	Grover's Algorithm [Anguita et al., 2003] Adiabatic Theorem [Neven et al., 2008] Hamiltonian Simulation [Rebentrost et al., 2014]	Quadratic [Anguita et al., 2003, Neven et al., 2008] Exponential [Rebentrost et al., 2014]
Bayesian Network	Adiabatic Theorem [O'Gorman et al., 2015]	Polynomial
Associative Memory	Quantum Memory [Ventura and Martinez, 2000] Grover's Algorithm [Ventura and Martinez, 2000] Hamiltonian Simulation [Trugenberger, 2001]	Quadratic [Ventura and Martinez, 2000]
Feedforward Artificial Neural Network	Quantum Measurement [Lewenstein, 1994] Adiabatic Optimization [Neigovzen, et al., 2009] Grover's Algorithm [Wiebe et al., 2014]	Quadratic [Wiebe et al., 2014]
Manifold Learning (e.g., Isomap)	Grover's Algorithm [Aïmeur et al., 2013]	Exponential
Boosting (e.g., AdaBoost)	Adiabatic Theorem [Neven et al., 2008]	Quadratic

CHAPTER 6

Quantum Filtering and Control

In classical robotics, filtering and control algorithms go together, often as software routines in a joint sensing and action loop. Filtering algorithms help infer a robot's state from noisy perceptual data of the environment that its sensors provide. Control algorithms utilize current state estimates to help facilitate the robot's trajectory.

Hidden Markov Models (HMMs) and Kalman Filters are two common classical filtering strategies that have served useful in smoothing time series of sensor data and correcting for statistical noise factors involved. PID (which stands for "Proportional-Derivative-Integral") is a classical control algorithm that helps facilitate a robot's actions, allowing for precise robot motion and correction for errors in execution.

An interesting research direction in quantum robotics is development of filtering and control algorithms that can operate on quantum phenomena as opposed to classical percepts. Sensing and control of quantum phenomena is a very different problem from the classical case. Directly observing an object governed by quantum dynamics runs the risk of changing the system's behavior, so direct sensing is generally not feasible. Control algorithms in quantum environments must take into account the unique nature of quantum mechanics in order to successfully manipulate quantum-scale objects. We will discuss approaches to sensing and controlling quantum phenomena.

6.1 QUANTUM MEASUREMENTS

Measuring and extracting information from a quantum system is necessary to implement quantum filtering and control algorithms. However, as thematically seen before, extracting information from a quantum system without collapsing it can be nontrivial. Two suggested ways in the literature of extracting information from a quantum system are projective and continuous measurements. In-depth background on quantum measurement can be found in Deutsch [1983], Gisin [1984].

6.1.1 PROJECTIVE MEASUREMENTS

In quantum mechanics, an observable physical quantity can be represented by a Hermitian operator in a Hilbert space. The Heisenberg uncertainty principle, however, suggests that two non-commutative observables cannot be simultaneously measured. For example, there are fundamental limits to the precision with which the position and momentum of a particle can be simultaneously known. A common measurement model to overcome some of challenges is projective measurements (also known as von Neumann measurements).

Because an observable M is an Hermitian operator, it can be represented as a weighted sum of orthogonal projectors acting on the state space . Thus, one can write:

$$M = \sum_i m_i P_i \tag{6.1}$$

where P_i is a projector onto the eigenspace M that has eigenvalue m_i. When a measurement occurs, eigenvalue m_i is returned with probability $p(m_i) = \langle \psi | P_i | \psi \rangle$, and the state of the system collapses to $\frac{P_m | \psi \rangle}{\sqrt{p(m)}}$.

Projective measurements are considered to be instantaneous, which is a reasonable assumption when the measurement signal is sufficiently strong and short-duration (the closer to an impulse, the better). In practice, it is not always feasible to build instantaneous quantum measurement methods. Thus, continuous measurements are often used to measure information from quantum systems.

6.1.2   CONTINUOUS MEASUREMENTS

Continuous measurements allow one to monitor an observable quantity over time. Continuous measurements monitor signals that appear on a longer-term time-scale, which is useful for extracting control feedback information [Jacobs and Steck, 2006].

An example of the continuous measurement methodology is letting the master quantum system of interest act upon an ancilla quantum system that can be observed using projective measurements. The system of interest can be transitively observed via the ancilla quantum system. While the ancilla quantum system is measured directly, the system of interest is not. Thus, the system of interest is allowed to evolve with minimal disruption of its quantum state. Hidden Quantum Markov Models (discussed in upcoming Section 6.2.2), Quantum Kalman Filters (discussed in upcoming Section 6.3.2), and the Stochastic Master Equation control strategy (discussed in upcoming Section 6.4.5) all leverage continuous measurement methodologies.

6.2   HIDDEN MARKOV MODELS AND QUANTUM EXTENSION

In a classical sensing scenario, a robot's sensors provide measurements that can be used to infer the robot-environment state. Reasoning over a time series of measurements in a dynamic environment can be a nontrivial statistical inference problem. The Hidden Markov Model (HMM) is a tool to help simplify this inference problem into one that is more computationally tractable.

6.2.1   CLASSICAL HIDDEN MARKOV MODELS

Hidden Markov Models (HMMs) are a method for representing probability distributions over data sequences. Imagine one has a time series of sensor data. In a robotics context, one may like to know the state of the robot and world from the sensor data. The true state of the system is "hidden" (i.e., unknown) at all times, though the sensor data received provides clues as to the time-evolution of the system. The HMM provides probabilistic reasoning capability to attempt to infer the state of the world from the time series of percept data.

Hidden Markov Model (HMM) Definition

- S, a set of possible system states.

- Y, a set of possible system observations.

- $T(s_i, s_j)$, a $|S| \times |S|$ matrix of transition probabilities which expresses the probability of a system evolving from state s_i to state s_j.

- $O(o_i, s_j)$, an observation model (a $|Y| \times |S|$ matrix) which expresses the probability of obtaining observation o_i in state s_j.

- p_0, a $|S| \times 1$ probability vector which expresses the distribution of system start states.

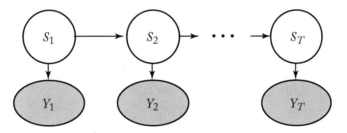

Figure 6.1: Graphical structure of Hidden Markov Model.

The graphical model of an HMM is shown in Figure 6.1. The graphical model of the HMM encodes the Markov assumption that knowledge of the current state means future events are independent of events prior to the current state:

$$P(s_{t+1}|s_t, s_{t-1}, \ldots, s_0) = P(s_{t+1}|s_t). \tag{6.2}$$

The evolution of the system state can thus be described by the sequential application of the transition matrix. Let s_n be the $|S| \times 1$ state distribution at a particular time n. Using the Markov assumption, it can be computed as:

$$s_n = (T)^n p_0 \tag{6.3}$$

where p_0 is the distribution of system start states. The probability of making observation a at time n is:

$$P(o_{a,n}) = O_a s_n \tag{6.4}$$

where O_a is the $1 \times |S|$ row vector corresponding to observation a in matrix O. Through sequential application of these equations, one can obtain the probability of a sequence of independent observations $o_{a,1}, o_{b,2}, o_{c,3}$:

$$
\begin{aligned}
P(o_{a,1}, o_{b,2}, o_{c,3}) &= P(o_{a,1}) P(o_{b,2}) P(o_{c,3}) \\
P(o_{a,1}, o_{b,2}, o_{c,3}) &= O_a s_1 O_b s_2 O_c s_3.
\end{aligned}
\tag{6.5}
$$

Though the true state of the system may be unknown at each time step, it can be inferred from the probabilities of obtaining particular observations in a given state based on the observation model, the distribution of initial system states, and the dynamics of the system expressed by the transition matrix. The Forward-Backward algorithm of the HMM provides not only the forward prediction described in Equations (6.3) and (6.4) but retrospective smoothing on state estimates upon input of latest data points.

An excellent treatment of HMMs is given by Rabiner and Juang [1986]. In robotics, HMMs have found a myriad of applications in sensor fault diagnosis [Verma et al., 2004], terrain mapping and classification [Wolf et al., 2005], understanding of human intent [Kelley et al., 2008], learning from human demonstration [Hovland et al., 1996], and robot introspection [Fox et al., 2006]. Given the widespread use of the classical HMM in representing various sensing and denoising scenarios in robotics, one may expect Hidden Quantum Markov Models to provide similarly significant contributions to the field of quantum robotics.

6.2.2   HIDDEN QUANTUM MARKOV MODELS

A Hidden Quantum Markov Model (HQMM) [Clark et al., 2015] is the quantum version of the HMM, and operates at quantum scale. The HQMM helps filter and make sense of a system governed by quantum dynamics rather than classical dynamics.

The HQMM starts with initial state probabilities of the quantum system that can be described by a density matrix, ρ_s. At each time step, the quantum system evolves and produces an output symbol. The HQMM is governed by a set of Kraus operations $\{K_m\}$. Recall from Section 4.3, a set of Kraus matrices $\{K_1, \ldots, K_\kappa\}$ satisfies $\sum_{i=1}^{\kappa} K_i^\dagger K_i = I_d$. Given this, the time evolution of the initial density matrix is given by:

$$\rho(t + \delta t) = \sum_{i=1}^{\kappa} K_i \rho_S(t) K_i^\dagger \tag{6.6}$$

where t is time and δt is a small change in time. The observation probabilities of the sequence of observations $o_{a,1}, o_{b,2}, o_{c,3}$ is given by:

$$p(o_{a,1}, o_{b,2}, o_{c,3}) = K_c K_b K_a \rho_S K_a^\dagger K_b^\dagger K_c^\dagger. \tag{6.7}$$

This rule is similar to the classical case in that sequential application of matrices yields the final result. The key difference is the Kraus matrices represent both the state transition matrix and observation matrix jointly. For the HQMM, these are not independent processes but very much intertwined.

A key ongoing research challenge is to find ways to implement Kraus operators for HQMMs. One potential way is to utilize an ancilla quantum system whose internal state interacts with the HQMM's qubits. The ancilla system's state is read out via projective measurements at each time step to provide indirect information about the primary quantum system. The primary quantum system's internal state remains hidden, though clues are given through its entanglement with this ancilla system.

Some proposed methods of implementing the overall HQMM include using the successive, non-adaptive readout of entangled many-body states [Monras et al., 2011], the time evolution of an open quantum system on a coarse-grained time scale [Sweke et al., 2014], and using an open quantum system with instantaneous feedback with interactions from the surrounding environment [Ralph, 2011].

Interestingly, the HQMM may be able to represent some stochastic processes more efficiently than HMMs [Monras et al., 2010]. The "realization problem" for HMMs [Vidyasagar, 2005] is: Suppose m is a positive integer and let $M = \{1, \ldots, m\}$. Suppose Y_t is a stationary process assuming values in M. Does there exist an HMM that reproduces the statistics of the process?

There exist some stochastic processes that have a more compact representation with HQMMs than HMMs. Monras et al. [2010] provides the following example of such a stochastic process. Consider the 4-symbol stochastic process ($s \in \{0, 1, 2, 3\}$) given by the transition matrices in Equation (6.8):

$$
T_0 = \begin{bmatrix} \frac{1}{2} & 0 & \frac{1}{4} & \frac{1}{4} \\ 0 & 0 & 0 & 0 \\ 0 & 0 & 0 & 0 \\ 0 & 0 & 0 & 0 \end{bmatrix} \quad
T_1 = \begin{bmatrix} 0 & 0 & 0 & 0 \\ 0 & \frac{1}{2} & \frac{1}{4} & \frac{1}{4} \\ 0 & 0 & 0 & 0 \\ 0 & 0 & 0 & 0 \end{bmatrix}
$$

$$
T_2 = \begin{bmatrix} 0 & 0 & 0 & 0 \\ 0 & 0 & 0 & 0 \\ \frac{1}{4} & \frac{1}{4} & \frac{1}{2} & 0 \\ 0 & 0 & 0 & 0 \end{bmatrix} \quad
T_3 = \begin{bmatrix} 0 & 0 & 0 & 0 \\ 0 & 0 & 0 & 0 \\ 0 & 0 & 0 & 0 \\ \frac{1}{4} & \frac{1}{4} & 0 & \frac{1}{2} \end{bmatrix}.
$$

$$(6.8)$$

Using a Hankel matrix bound [Anderson, 1999], [Monras et al., 2010] proves that the stochastic process cannot be represented by a two-state HMM and requires three states. However, the stochastic process can be represented by a HQMM with two states.

6.3   KALMAN FILTERING AND QUANTUM EXTENSION

A robot's sensors allow it to measure and quantify phenomena in its environment. Some of what is sensed is actual useful signal from sources of interest in the world. Much of what is sensed, however, is noise due to external factors such as the background. The goal of Kalman filtering [Kalman, 1960] is to improve resolution of signal from noise. Kalman filtering combines information from multiple sensor observations over time to improve measurement precision from percepts containing statistical noise. This section will discuss the details of the Kalman filter as well as how this tool can be extended to quantum (rather than classical) sensing scenarios.

6.3.1   CLASSIC KALMAN FILTERING

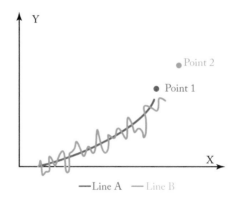

Figure 6.2: Kalman Filter can help infer true underlying system state from multiple noisy sensor observations.

Imagine a signal starting at time $t = 0$ and traveling to points 1 and 2 (Figure 6.2). Even though the signal has a clear trajectory, the underlying sensor data indicates that the trajectory is noisy. The purpose of the Kalman Filter is to improve estimation of the true state of the system based on combining information from multiple (potentially noisy) measurements and previous system state estimates. Kalman filtering is particularly common in navigation and control of mobile robots because of the noisy measurements that are obtained in these contexts that make state estimation of the robot challenging.

Mathematically, the Kalman Filter is described by the following equations:

$$x_k = Ax_{k-1} + Bu_k + w_{k-1}$$
$$z_k = Hx_k + v_k. \tag{6.9}$$

The first equation describes evolution of the system state, while the second equation describes the process by which measurements are created. The system state at time k is denoted as x_k. The matrix A is the state transition model that describes how the state evolves from time step to time step, B is the control-input model applied to an input control signal u_k, $w_{k-1} \sim N(0, Q_k)$ is the process noise model using a zero mean normal distribution with covariance Q_k, z_k is the received measurement, H is an observation model that maps the state space onto the measurement space, and $v_k \sim N(0, R_k)$ is a zero mean Gaussian white noise variable that represents measurement noise.

The following example illustrates the concept of the Kalman Filter. Consider a scenario where measurement yields a constant signal, there is no control input, but state value and noise are still measured. These conditions imply that $A = 1$, $u_k = 0$, and $H = 1$. Simplifying the previous equation set yields:

$$x_k = x_{k-1} + w_k$$
$$z_k = x_k + v_k. \qquad (6.10)$$

From these equations, one can see that both the state estimate x_k and measurement z_k consist of some state plus sensor noise. During each iteration of the filter, the current state estimate $\hat{x}_k$ is updated using:

$$\hat{x}_k = K_k z_k + (1 - K_k)\hat{x}_{k-1} \qquad (6.11)$$

where k represents the current time (typically a discrete-time step), K_k represents the Kalman gain, z_k represents the measured value, and $\hat{x}_{k-1}$ represents the previous state estimate.

The current state estimate is a fusion of the current measured value and the previous estimate. The Kalman gain K_k is an averaging factor that is computed using a covariance matrix describing state variability.

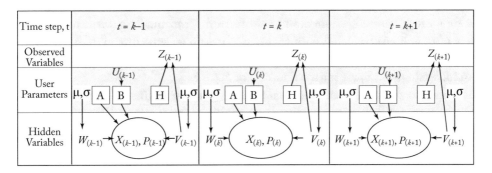

Figure 6.3: Graphical model structure of Kalman Filter.

The graphical model structure of a Kalman Filter (shown in Figure 6.3) describes the relationship between the filter parameters and demonstrates the alternating prediction (time update) and update (measurement correction) steps. P_k is the posterior error covariance matrix which indicates the estimated state accuracy and is used to compute the Kalman gain.

6.3.2   QUANTUM KALMAN FILTERING

Since measurement of a quantum system can cause it to decohere, gaining knowledge via direct measurement is generally infeasible. The Quantum Kalman Filter [Ruppert, 2012] utilizes "non-demolition" measurements, a continuous measurement methodology. Non-demolition measurements allow indirect information acquisition of a quantum system S by coupling it with an ancilla quantum system, M, that interacts with S. Directly measuring quantum system M provides information about S, without decoherence of S.

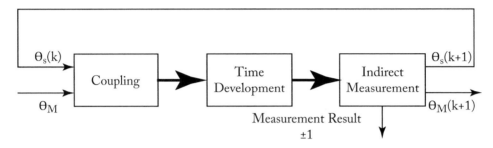

Figure 6.4: System diagram of quantum Kalman Filter. (Reprinted from Process Control Research Group, retrieved from `http://daedalus.scl.sztaki.hu`, by Ruppert [2012]. Reprinted with permission.)

The operational pipeline for the Quantum Kalman Filter is shown in Figure 6.4 for a single discrete time step. At the beginning of each time step k, an ancilla quantum system is prepared with a known state θ_M. The ancilla system M is coupled with the quantum system S, whose state θ_S is unknown. The two quantum systems are then allowed to evolve according to their joint dynamics for a preset sampling time duration. Afterward, the ancilla qubit is measured and the qubit's new state will be correlated with that of the unknown quantum system. By running the filter for multiple time steps and aggregating information across multiple measurements, one can achieve an increasingly accurate state estimation of S.

The state and measurement equations of the Quantum Kalman Filter provide the mathematical machinery for fusing information from multiple observations. The equations are analogous to their classical counterparts. However, because the measurement process is indirect as opposed to direct, the quantum and classical filters are quite different mathematical objects. The Quantum Kalman Filter equation system is given by:

$$
\begin{aligned}
x_{k+1} &= x_k + Nc^2 x_k \left(1 - x_k^2\right) + \omega_k c \left(1 - x_k^2\right) \\
y_k &= Ncx_k + \omega_k \\
\hat{x}_{k+1} &= \hat{x}_k + Nc^2 \hat{x}_k \left(1 - \hat{x}_k^2\right) + K_k \left(y_k - Nc\hat{x}_k\right) \left(1 - \hat{x}_k^2\right)
\end{aligned}
\tag{6.12}
$$

where x_k is the current state, $\hat{x}_k$ is the current state estimate, y_k is the measurement, ω_k is the measurement noise, N is the number of time steps, c is a parameter that characterizes the state of the ancilla qubit and its interaction with the qubit being monitored, and K_k is the Kalman gain (whose computation is extensively described in Ruppert [2012]). Ruppert [2012] also provides a sensitivity analysis of parameters of the Quantum Kalman Filter in simulation.

One of the key areas of exploration for moving from theory to practice is quantifying what conditions and requirements constitute a viable ancillary system to monitor a master quantum system. Ideally, the ancilla system should be able to provide sufficient information about the quantum system of interest through coupling, though must not significantly alter the state of the original system or its natural evolution. Much progress must still be made to develop practical Quantum Kalman Filters. One of the most useful applications of the technology could be in quantum control, which is discussed next.

6.4 CLASSICAL AND QUANTUM CONTROL

The study of controlling quantum phenomena has been a concurrent goal of quantum mechanics alongside much of quantum physics and chemistry. In robotics, a primary goal is to develop methods for active manipulation and feedback control of quantum systems. The nuances of entanglement and coherence make control of quantum systems more challenging than their classical counterparts. In particular, classical models are not transferable to quantum systems because both sensing and actuation behave differently. Thus, new techniques are required. Quantum control methods could extend the reach of robotics to operate in quantum-scale environments and to manipulate and control quantum phenomena. Some of the key results in classical and quantum control are discussed in this section. Dong and Petersen [2010] provide a comprehensive survey of quantum control theory and applications, and their paper is suggested for further reading.

6.4.1 OVERVIEW OF CLASSICAL CONTROL

There are two major types of classical control approaches: open loop and closed loop. In open loop control, the controller commands are computed using only system state and a state transition model, with no additional input. Open loop systems do not incorporate nor compensate for errors during execution of the control commands. Closed loop (or feedback control) systems, in contrast, incorporate measurement information into the controller and attempt to drive the system to a specified desired state. Closed loop systems compute an error, often based on the difference between a current state and the system settling point, and use this error to adjust inputs to help drive a system to the desired system state.

Proportional-Integral-Derivative (PID) control is a common strategy for implementing closed loop robot systems [Siciliano et al., 2010]. The mathematical form for PID is:

$$u(t) = K_p e(t) + K_i \int_0^t e(t)dt + K_d \frac{de(t)}{dt} \tag{6.13}$$

where $u(t)$ is the time-varying controller input and $e(t)$ is the error signal (defined as the desired point minus the measurement point). There are three gains, one for each mathematical term in the PID equation, that give rise to PID's namesake. K_p is the proportional gain, K_i is the integral gain, and K_d is the derivative gain.

The proportional mathematical term specifies how quickly the controller corrects for deviations from the desired point. The correction factor in the control is in proportion to the error. Larger time-varying errors given by $e(t)$, the error signal, will generally receive larger correction with the next control signal. The proportional gain K_p specifies how much of a correction factor is applied per time step.

If one uses solely the proportional term in the controller, the input signal may oscillate about the desired state, alternating between over-compensating and then under-compensating for the desired system state. The derivative term specifies a damping factor that, over time, smooths such oscillations toward the desired state. The K_d gain specifies how much damping is applied.

On their own, the previous two mathematical terms are not enough for the system to guarantee reaching the desired state. Due to errors that accumulate over time while the controller is trying to correct itself, the controller may converge to a steady state too quickly that is not the desired state. To propel the system state to the desired state, the integral term accounts for cumulative error, and is gated by the K_i gain. In additive combination, these three mathematical terms yield a sufficiently robust control strategy for many applications.

6.4.2   OVERVIEW OF QUANTUM CONTROL MODELS

Quantum control models break down across several dimensions. One dimension is whether the quantum system being controlled is open or closed. Open quantum systems interact with their external environment. Closed quantum systems, in contrast, do not. Another dimension is whether feedback is provided by the measurement process or not. Finally, there are two major views of quantum mechanics, the Schrödinger picture and the Heisenberg picture, under which quantum control algorithms can be formulated. In the Schrödinger picture (as previously seen with the Schrödinger equation in Section 2.3), the state vector of the quantum system $|\psi\rangle$ evolves in time but the operators and observables are generally constant with respect to time. The Heisenberg picture takes the opposite view in that the state vector $|\psi\rangle$ does not change with time but the observables do.

In upcoming sections, we discuss quantum control methods that apply to these different types of control problems. Bilinear Model (BLM) approaches are designed to control closed quantum systems. Markovian Master Equation (MME) approaches are designed to control open quantum systems. Stochastic Master Equation (SME) approaches are designed for quantum feedback control. All three of these models use the Schrödinger model for quantum mechanics. Linear Quantum Stochastic Differential Equation (LQSDE) approaches are designed to control quantum systems that use the Heisenberg formulation. Jointly, these methods allow extending classical control ideas to quantum control.

6.4.3   BILINEAR MODELS (BLM)

Bilinear Models (BLMs) are used to control closed quantum systems. A BLM expresses the state equations for system control in the form $y^T A_i x = g_i$ where A_i are matrices and g_i are real numbers.

Recall from Chapter 2 that in a closed quantum system, the quantum state $|\psi(t)\rangle$ evolves as described by the Schrödinger equation: $i\hbar \frac{\delta}{\delta t}|\psi\rangle = H|\psi\rangle$ (where $i = \sqrt{-1}$). The goal is to find a final time $t > 0$ and control inputs $u_k(t) \in \mathbb{R}$ which drive the quantum system from its initial state $|\psi_0\rangle$ to a target state $|\psi_f\rangle$. Define the total Hamiltonian acting on the system as $H(t) = H_0 + \sum_k u_k(t)H_k$ and the unitary transformation $U(t)$ that describes the quantum state as a function of time: $|\psi_t\rangle = U(t)|\psi_0\rangle$. Plugging these definitions into the Schödinger equation, one obtains:

$$i\dot{U}(t) = H(t)U(t), \ U(0) = I$$
$$H(t) = H_0 + \sum_k u_k(t)H_k. \tag{6.14}$$

The Hamiltonian $H(t)$ describes the controlled evolution of the system given its time-varying control inputs $u_k(t)$. Solving these equations leads to a controller for the closed quantum system.

BLMs have many application areas. They are typically utilized to describe closed quantum control systems such as molecular systems in physical chemistry and spin systems in nuclear magnetic resonance [Alessandro and Dahleh, 2001].

6.4.4   MARKOVIAN MASTER EQUATION (MME)

Not all quantum systems are closed. In many applications, the quantum system is open and thus unavoidably interacts with its external environment. One way of modeling these types of systems is to use Markovian Master Equations (MME).

Consider an arbitrary quantum system that can take finite discrete states described by a time-varying density matrix $\rho(t)$. The time evolution of the system can thus be expressed using what is referred to as a Master equation for the system:

$$\dot{\rho}(t) = A(t)\rho(t). \tag{6.15}$$

When the state transition matrix A is time-independent (i.e., the Markov assumption), Equation (6.15) can be simplified to $\dot{\rho}(t) = A\rho(t)$. A commutation operator, defined as $[X, \rho] = X\rho - \rho X$, can be used to write the quantum Liouville equation:

$$i\dot{\rho}(t) = [H(t), \rho(t)] \tag{6.16}$$

where $i = \sqrt{-1}$. In an open quantum system, the overall system-environment Hamiltonian can be written as:

$$H = H_S + H_E + H_{SE} \tag{6.17}$$

where H_S is the Hamiltonian of the quantum system, H_E is the Hamiltonian of the external environment, and H_{SE} is the interaction between the system and the environment. The MME strategy is to treat the interaction term H_{SE} as a perturbation of the original state of the system and environment $H_0 = H_S + H_E$. Define the time-varying interaction equation:

$$\tilde{H}_{SE}(t) = e^{iH_0 t} H_{SE} e^{-iH_0 t}. \tag{6.18}$$

Treating the combined system (quantum system + environment) as closed, the Liouville equation can be written as:

$$i\dot{\tilde{\rho}}(t) = \left[\tilde{H}_{SE}(t), \tilde{\rho}(t)\right]. \tag{6.19}$$

The formal solution is:

$$\tilde{\rho}(t) = \rho(0) - i \int_0^t ds \left[\tilde{H}_{SE}(s), \tilde{\rho}(s)\right]. \tag{6.20}$$

MME can be applied when a system has a short environmental correlation time, and negligible memory effects. Unlike BLMs, MMEs can be used for open quantum systems as opposed to just closed quantum systems. MMEs have been applied to quantum error correction [Ahn et al., 2002] and spin squeezing [Thomsen et al., 2002].

6.4.5   STOCHASTIC MASTER EQUATION (SME)

Stochastic Master Equations (SMEs) are a tool to develop quantum feedback controllers that can utilize measurement information from a quantum system to robustify control of the system. Let $\rho(t)$ be the time-varying density matrix of the quantum system and H the Hamiltonian that acts on the system. Assuming a continuous measurement scheme is used to measure an observable X of the system, the evolution of the quantum system's density matrix can be described by:

$$d\rho = -i[H, \rho]dt - \kappa[X, [X, \rho]]dt + \sqrt{2\kappa}(X\rho + \rho X - 2\langle X\rangle \rho)dW \tag{6.21}$$

where κ is a parameter related to measurement strength, $\langle X\rangle = tr(X\rho)$, and dW is a Wiener increment with zero mean and unit variance equal to dt and satisfies:

$$dW = dy - 2\sqrt{\kappa}\langle X\rangle dt \tag{6.22}$$

where y is the measurement value. The role of the Wiener increment is to incorporate new measurement information. As the measurement value changes over time, the Wiener increment changes and is incorporated into the evolution of the overall quantum system.

In general, many types of SMEs exist, and the form which the master equation takes depends on the measurement process that is being modeled. The SME is useful for modeling atomic interaction with electromagnetic fields and controlling quantum optical systems [van Handel et al., 2005].

6.4.6 LINEAR QUANTUM STOCHASTIC DIFFERENTIAL EQUATION (LQSDE)

The previously discussed models all use the Schrödinger interpretation of quantum mechanics where equations describing the time-dependence of quantum states are given. In contrast, the Linear Quantum Stochastic Differential Equation (LQSDE) approach uses the Heisenberg interpretation, which can be more convenient when time-dependent operators on $\mathcal{H}$ describe the quantum dynamics. In the Heisenberg interpretation of quantum mechanics, the operators acting on the system evolve in time instead of quantum states.

The general form for the LQSDE is:

$$
\begin{aligned}
dx(t) &= Ax(t)dt + Bdw(t) \\
dy(t) &= Cx(t)dt + Ddw(t)
\end{aligned}
\tag{6.23}
$$

where $x(t)$ is a vector of time-varying self-adjoint[1] possibly noncommutative system operators. A, B, C, and D are appropriately sized matrices specifying the system model. The initial conditions of the system, $x(0) = x_0$ are operators that satisfy the following commutation relations:

$$
[x_j(0), x_k(0)] = 2i\,\Theta_{jk}, \; j, k = 1, \ldots, n
\tag{6.24}
$$

where Θ is a real antisymmetric matrix. In the LQSDE model, the input signal can be decomposed as:

$$
dw(t) = \beta_w(t)dt + \tilde{w}(t)
\tag{6.25}
$$

where $\tilde{w}(t)$ is quantum noise and $\beta_w(t)$ is a self-adjoint, adapted[2] process [Parthasarathy, 2012].

LQSDEs are useful because they describe some non-commutative linear stochastic systems, especially in linear quantum optics [James et al., 2007] and nanoscale technology [Zhang et al., 2009].

6.4.7 VERIFICATION OF QUANTUM CONTROL ALGORITHMS

When designing classical robot controllers, analysis is often done to understand key properties of the control algorithm such as its controllability, stabilizability, and reachability [Siciliano et al., 2010]. Such verification analysis helps ensure a robot controller performs within its intended specification. Similar verification analysis is likely necessary for quantum control algorithms, though the probabilistic nature of quantum mechanics makes analysis of controller properties more challenging than classical control. Model checking of quantum systems is a developing area of research [Ying et al., 2014, Ying and Ying, 2014].

---

[1]An operator A is self-adjoint if $A = A^{\dagger}$.
[2]For our purposes, this just means $\beta_w(t)$ is known at time t (though not necessarily before time t).

6.5   CHAPTER SUMMARY

<u>**Chapter Key Points**</u>

- Hidden Markov Models represent probability distributions over time series of sensor data. Hidden Quantum Markov Models (HQMMs) are formulated similarly, but the Kraus matrices encapsulate observation and transition information jointly.

- Quantum Kalman Filtering extends the Kalman filter to filtering on systems governed by quantum dynamics. Indirect measurements extract information from a quantum system without disrupting its superposition.

- Quantum control methods extend classical control models (summarized in Table 6.1).

Table 6.1: Summary of quantum operating principles discussed in quantum filtering and control

Technique	Quantum Operating Principles (QOPS) Used	Potential Advantages (over classical version)
Hidden Quantum Markov Models [Clark et al., 2015]	Quantum Measurement	More compact representation for stochastic processes
Quantum Kalman Filtering [Ruppert, 2012]	Quantum Measurement	Capability to filter quantum system
Bilinear Models	Quantum State Evolution	Capability to control closed quantum systems
Markovian Master Equation	Quantum State Evolution	Capability to control open quantum systems
Stochastic Master Equation	Quantum State Evolution	Extension of feedback control to quantum systems
Linear Quantum Stochastic Differential Equation	Heisenberg Picture of Quantum Mechanics	Capability to control quantum systems using the Heisenberg model

CHAPTER 7

Current Strategies for Quantum Implementation

In this chapter, we discuss recent efforts to implement quantum engineering systems, with an emphasis on quantum computer development efforts. A working quantum computer is essential to obtain many of the speedups discussed in previous chapters such as quantum search (Chapter 3), quantum agent planning (Chapter 4), and quantum machine learning (Chapter 5). Additional breakthroughs in the actual implementations of quantum mechanisms are necessary for quantum robotics to attain its full potential. We begin with a theoretical discussion of two frameworks used to evaluate quantum computers: the DiVincenzo definition and the Mosca classification. We then highlight a range of experiential approaches which researchers have taken to develop a working quantum computer, and conclude with a discussion of D-Wave, to date the most commercially successful of these approaches.

7.1 DIVINCENZO DEFINITION

One of the first frameworks developed in quantum computing was developed by David DiVincenzo. He laid out a series of requirements for the quantum circuit model of computation to be implemented [DiVincenzo, 2000]. Each of these requirements are necessary for a system based on quantum gates to utilize quantum effects for computation, but even taken together, they are insufficient for a fully functional quantum computer. More work remains to identify the specific physical and practical requirements of a quantum computer. The basic requirements laid out are as follows.

First, a physical system that contains qubits must exist. A particle is classified as a qubit if its state can fill a two-dimensional complex vector space. Furthermore, the qubit ought to be "well-characterized"—that is, the qubit's internal Hamiltonian must be known, along with how it interacts with other qubits and its environment. And critically, the qubits in the system must exhibit entanglement.

Second, the system must have the capability of inducing qubits to initialize to a simple fiducial state—for example, $|00\rangle$. This can be implemented by having the system rest until it returns to the ground state of its Hamiltonian or by forcing a series of measurements upon the system which induce the desired state. An important practical consideration is the length of time required for this reset to occur.

Third, the qubits' decoherence times must be significantly longer than the gate operation time. The window of opportunity to experience quantum computational speedups is very narrow. Energized qubits naturally have very short decoherence times, after which they exhibit standard classical behavior. Beyond this point the speedups and other advantages of using a quantum machine will no longer exist. This is one of the significant unsolved physical challenges facing the field today [Ladd et al., 2010]. Two approaches are currently being considered to tackle this challenge. One is directly lengthening the decoherence time of a system's qubits, or more precisely, experimenting with new materials to construct qubits whose decoherence times are longer than current experimentally measured results. The second approach is to implement more robust error correction functionality into the system. For example, ancillary qubits may be used to detect decoherence, and an appropriate correction adjustment can be made to the system.

Fourth, the system should be able to exhibit full control over a set of universal quantum gates. Such a set would enable the system to perform any computation possible for a quantum computer. This has proven difficult in practice. All physical implementations to date have only been able to demonstrate a limited range of Hamiltonian transformations. Workarounds can make such a quantum computer usable in specific circumstances. "A First gen" quantum system will likely be built optimized toward one specific function—more calculator than computer. It is possible quantum systems may become more general in the future.

Fifth, the system must have the capability to read individual qubits to generate the output. Since qubits rest in superposition, the output itself will often be probabilistic. That is, the qubit may output the correct response only a fraction of the time. To compensate, repeated system runs may need to be conducted and measured, to obtain a reliable estimator of the correct output.

Two additional requirements exist if the quantum system is to communicate with other systems, a greatly beneficial feature in the realm of quantum robotics. The first is the ability to convert stationary qubits embedded in the system into "flying qubits" optimized for transportation (or otherwise transmit information between these two types of qubits). The second is the ability to transport the flying qubits from one location to another.

Table 7.1: DiVincenzo definition of a quantum computer

DiVincenzo Definition
Criteria 1. A physical system that contains qubits must exist
Criteria 2. Capability to induce qubits to initialize
Criteria 3. Decoherence time much longer than gate operation time
Criteria 4. Full control over set of universal quantum gates
Criteria 5. Capability to read individual qubits
Criteria 6. (Communication) Ability to convert stationary qubits into flying qubits
Criteria 7. (Communication) Ability to transport flying qubits from one location to another

7.2   MOSCA CLASSIFICATION

The Mosca Classification [Mosca et al., 2014] provides a scale of five increasingly restrictive categories for determining whether a computer can be classified as quantum. Level 1 is the least restrictive category, and Level 5 is the most restricive. A computer classified at a particular level meets all the requirements of the previous levels. A Level 3 machine, for example, meets the requirements stated in Levels 1, 2, and 3.

Table 7.2: Mosca classification of a quantum computer

Mosca Classification
Level 1. Since the world is quantum, any computer is by construction a quantum computer
Level 2. A quantum computer is a computer that uses intrinsically quantum effects that cannot naturally be modeled by classical physics
Level 3. A quantum computer is a computer that uses intrinsically quantum effects to gain some advantage over the best known classical algorithms for some problem
Level 4. A quantum computer is a computer that uses intrinsically quantum effects to gain an asymptotic speed-up over the best known classical algorithms for some problem
Level 5. A quantum computer is a computer that is able to capture the full computational power of quantum mechanics, just as conventional computers are believed to capture the full computational power of classical physics

At Level 1, Mosca defines any computer to be a quantum computer since the world we inhabit is a quantum world. That is, all computers, including conventional ones, run governed by the laws of quantum physics. In the case of most computers today, the quantum effects aggregate into classical effects, so the phenomena which exist only in the quantum realm (e.g., entanglement or superposition) remain unobserved and do not impact their performance. A computer which does explicitly leverage quantum phenomena like entanglement or superposition in its operation, would meet the requirements of Level 2 and thus could be categorized as at least a Level 2 Mosca machine.

A computer could attain a Level 3 classification if the quantum effects it leverages enable the computer to gain an advantage in speed or performance over the best known classical algorithms for a particular problem. Level 4 is reached when a computer is not only able to perform slightly better, but demonstrates an asymptotic speed-up over the best known classical algorithms for a particular problem. It is important to note that both of these categories allow for specialized quantum computers, which are optimized toward performing specific functions. Such computers need not be deployable toward the full spectrum of computational applications in the way that we understand the most powerful classical computers operating today. For example, it would be sufficient for a Level 4 quantum computer only to be able to apply Shor's algorithm to perform prime factorization asymptotically faster than any classical computer to date for it to maintain its categorization. It would not be necessary for the computer to be a general purpose machine, deployable in a wide array of computations. Finally, a computer which is able to capture the full computational power of quantum mechanics would be categorized as a Level 5 Mosca machine, the highest and most restrictive category in the Mosca Classification.

7.3 COMPARISON OF DIVINCENZO AND MOSCA APPROACHES

The DiVincenzo and Mosca frameworks approach the classification of quantum computers from different angles. The DiVincenzo takes a circuit-based approach which builds up to a quantum computer; the Mosca evaluates potential quantum computers by analyzing the algorithms they compute.

Additionally, the DiVincenzo framework provides a simple binary evaluation of a computing machine. Under the DiVincenzo framework, a computer may be classified as either a universal quantum computer or not at all. In contrast, the Mosca framework provides five increasingly restrictive categories for a quantum computer. (A Level 5 Mosca classification is broadly equivalent to meeting the standards of the DiVincenzo framework.) No quantum implementations to date are able to meet either the DiVincenzo definition or the Level 5 Mosca classification for a quantum computer. Therefore at this time, the Mosca is the more useful framework for evaluating quantum implementations and the one we shall apply later in this chapter.

7.4 QUANTUM COMPUTING PHYSICAL IMPLEMENTATIONS

A number of physical implementation schemes have been proposed by researchers around the world. This section will introduce a selection of approaches, to highlight the range of options being considered and the lack of clarity surrounding which approach is ideal for implementing a quantum computer.

The ion trap quantum computer is one of the earlier experimental implementation approaches [Cirac and Zoller, 2000, Steane, 1997b]. It works by using positively charged ions, or cations, to represent qubits. Ions are commonly generated from an alkaline earth metal [Steane, 1997a], but other elemental candidates have also been used [Kielpinski et al., 2002]. The approach is to create ions and then capture them (for example, by firing a beam of high energy electrons through a cloud of the element chosen, then capturing them using charged electrodes in a vacuum). A single ion represents a qubit based on its spin. Researchers can use lasers to control that spin for initialization and during computation.

A related approach is to represent qubits not with ions but with neutral atoms [Briegel et al., 2000]. The atoms' lack of charge is both the method's greatest strength and weakness when compared to ion trap [Hughes et al., 2004]. Since the atomic qubits are not charged, they are less likely to interact with the environment and experience decoherence. However, this neutrality means they will be less likely to interact with each other, decreasing the desired entanglement of a quantum system. Thus, entanglement has to be induced, for example by positioning atoms very close to each other [Soderberg et al., 2009]. The challenge is whether physically scalable approaches exist, such that atomic qubits interact sufficiently with each other while avoiding decoherence long enough to complete quantum computation.

Photons have also been used as part of experimental quantum systems. Optical quantum computing represents qubits using photons, which has the advantage of smaller particle size (resulting in higher potential speeds) and lower decoherence rates compared to ions or atoms [Kok et al., 2007]. Cavity Quantum Electrodynamics (QED) takes a hybrid approach, interacting trapped atoms with photon fields to generate entanglement [Pellizzari et al., 1995].

In yet another approach, electrons can be contained within quantum dots that trap individual electrons and utilize them as qubits [Lent et al., 1994, Loss and DiVincenzo, 1998]. Behrman et al. [1996] utilizes this approach to implement Feynman path integrals, which were discussed in Equation (5.38). Quantum dot molecules are groups of atoms placed in spatial proximity to each other. The nearness of the atoms allows electrons to tunnel between the dots and create dipoles. Optical tools can be used to control the configuration of electron excitations, leading to a trainable artificial neural network. A temporal artificial neural network can be created where the nodes in the network correspond to system evolution time slices, and the spatial dot configuration allows multi-layer networks to be trained. The quantum dots approach has been extended by Altaisky et al. [2015], Tóth et al. [2000].

The approach favored by both IBM [Córcoles et al., 2015] and D-Wave [Bunyk et al., 2014] is to leverage quantum behavior which exhibits when materials experience superconductivity. Superconductivity is a physical state achieved at a specific temperature threshold, at which zero electrical resistance is experienced and magnetic fields are expelled. The exact threshold depends on the chemical composition of the material being super-conducted, and is often near zero Kelvin. The near-absence of nearly all heat, electrical resistance, and electromagnetic interference make this environment a potentially strong candidate for quantum computing.

The different physical implementations discussed above lend themselves to different algorithmic designs. For example, D-Wave has had success with Bayesian network structure learning (Section 5.6.2), while Altaisky [2001] has demonstrated an optical approach to quantum artificial neural networks (Section 5.6.2).

To highlight the progress within the field thus far, the next section will discuss one of these implementations in further detail: the D-Wave machine. It will provided a detailed review of the literature examining the quantum effects demonstrated by the D-Wave and its algorithmic performance against various classical benchmarks.

7.5   CASE STUDY EVALUATION OF D-WAVE MACHINE

D-Wave's implementation has generated discourse in recent years, not only within the research community but also in the general public. Observers debate the significance of the experimental results run on D-Wave machines, and are even divided on the question of whether the D-Wave machine can even be considered a quantum computer. Part of the confusion stems from inconsistent criteria for qualifying a device as a quantum computer (e.g., What characteristics qualify a computer as quantum? To what extent does computing speed impact this qualification?) To address this discussion, we apply the Mosca Classification to selected publication results analyzing D-Wave and its performance. While a comprehensive overview of all papers published on D-Wave over the last two decades is beyond the scope of this book, we hope to provide a starting point for delving into the debate. We encourage the reader to build upon our work by applying the framework to other publications on D-Wave, and other quantum computing implementations. Doing so makes it possible to arrive at a nuanced evaluation of the progress a proposed quantum computer has achieved thus far, and the remaining progress necessary to attain a universal quantum computer.

Analyzing D-Wave (or any implemented quantum computer) under the first criteria yields no debate—any physically constructed computer automatically qualifies as a quantum computer. Examining D-Wave under the latter criterion yields further discussion. The remainder of this section will survey the literature and analyze the extent to which D-Wave qualifies as a quantum computer. It will find that D-Wave clearly qualifies at least as a Level 2 Mosca machine, and is making progress toward entering Level 3.

Johnson et al. [2011] noted that classical thermal annealing should exhibit linear correlation between time and temperature while quantum annealing would not. They tested an 8-qubit D-Wave system and found no linear correlation, which suggests the presence of quantum activity. Lanting et al. [2014] also found evidence of quantum activity in the 108-qubit D-Wave One using a method called qubit tunneling spectroscopy (QTS), which measures the eigenspectrum and level occupation of a system. Tests provided evidence that the system demonstrated entanglement, a key feature of a processor running on intrinsically quantum effects. Boixo et al. [2013] noted that increasing temperatures would increase the speed of isolating a Hamiltonian solution in simulated annealing, but decrease the speed of the same process in quantum annealing. Their experiment observed evidence of quantum annealing.

Boixo et al. [2014] later simulated three processes: classical annealing, quantum annealing, and classical spin dynamics. Across all three, the study measured the success rate across a range of energy inputs and instances, then compared this simulated data with experimental data observed using the D-Wave One. It found the experimental data most closely correlated with the simulated quantum annealing. In response to this study, Smolin and Smith [2013] proposed two separate classical models that might also correlate with the experimental data. This presented a counter-hypothesis which suggested that the experimental data might not have resulted from a quantum process, but could have arrived from classical means. The research team of Boixo et al. [2014]

Table 7.3: Is D-Wave a quantum computer?

Authors	Computer	Publication Findings (Experiment Summary)
[Johnson et al., 2011] (D-Wave co-authored)	8 qubit	Quantum behavior detected (no linear correlation detected)
[Lanting, et al., 2014] (D-Wave co-authored)	D-Wave One	Quantum behavior detected (evidence of entanglement)
[Boixo et al., 2013]	D-Wave One	Quantum behavior detected (temperature decrease detected)
[Boixo et al., 2014]	D-Wave One	Quantum behavior detected (closest correlation to quantum annealing)
[Smolin and Smith, 2013]	D-Wave One	No quantum behavior detected (could be explained by classical means)
[Wang et al., 2013]	D-Wave Two	Quantum behavior detected (probabilistic evidence of superposition)
[Albash et al., 2015]	D-Wave One	Quantum behavior detected (matches Hamiltonian signature)

responded in Wang et al. [2013] and confirmed that the methods used in Smolin and Smith [2013] are valid, and that classical models could explain some of the results. However, they posit that when all the results of the previous paper are considered together, there is strong probabilistic evidence of superposition, and therefore quantum activity within D-Wave. Albash et al. [2015] builds upon this conclusion further by applying a generalized quantum signature Hamiltonian test to the D-Wave One, and again found evidence of quantum activity. Taken together, these studies collectively suggest that the D-Wave demonstrates strong evidence of quantum activity during its computation process. The reader is encouraged to build upon our findings, by applying the Mosca framework to quantum computing publications beyond those included here.

A further discussion surrounding D-Wave is over whether the system is truly faster than classical computing implementations and, if it is, whether those speedups are due to the quantum effects of superposition and entanglement.

In studies which compared D-Wave against classical algorithm implementations, the results were mixed. McGeoch and Wang [2013] tested the D-Wave One against conventional software solvers (CPLEX, TABU, and Akmaxsat) on estimating the solution to several NP-hard computational problems (QUBO, W2SAT, and QAP). This test found D-Wave performed 3600 times faster than its competitors, but the generalizability of the results is uncertain. In King et al. [2015], D-Wave pit its 2X system against two classical software solvers (simulated annealing and the Hamze-de Freitas-Selby algorithm) for Chimera-structured Ising problems within three categories: RANr, ACk-odd, and FLr. The three entrants were broadly competitive with one another, with D-Wave outperforming HFS while simulated annealing recorded the fastest times within the range of problem sizes tested. Rønnow et al. [2014] randomly selected problems and tested the 503-qubit D-Wave Two using a benchmark of random spin glass instances. They found no evidence of quantum speedup. Similarly Hen et al. [2015] compared the quantum performance of D-Wave Two against classical algorithms implemented specifically to compete against quantum counterparts. They also did not detect any quantum speedup. Most recently, Denchev et al. [2015] find that D-Wave 2X is able to achieve an asymptotic speedup (of 10^8 times) against a standard single-core simulated annealer and a constant speedup over Quantum Monte Carlo when tested against a specific class of problem. However, several classical algorithm variants were able to match the D-Wave in performance.

Thus, it seems that while D-Wave may outperform some classical algorithms in specific instances, it does not exhibit a significant advantage over all algorithms across all possible problems. For further information, see McGeoch [2012] for further discussion of proposed applications for D-Wave and other research conducted to understand its quantum properties. Though quantum computers do not outperform their classical equivalents today, it is quite possible that an upcoming technological advancement tomorrow could yield a Class 3 Mosca quantum computer.

Table 7.4: Does D-Wave run faster than classical algorithm implementations?

Authors	Computer	Publication Findings
[McGeoch and Wang, 2013]	D-Wave One	Quantum 3,600 times faster
[King, et al., 2015]	D-Wave Two	Mixed results in computation time comparisons
[Rønnow et al., 2014]	D-Wave Two	No quantum speedup detected
[Hen et al., 2015]	D-Wave Two	No quantum speedup detected
[Denchev et al., 2015]	D-Wave Two	10^8 times faster than some classical algorithms, no observable speedup compared to others

7.6 TOWARD GENERAL PURPOSE QUANTUM COMPUTING AND ROBOTICS

The quantum implementations discussed in this chapter, and all quantum implementations proposed to date, are limited-purpose specialty devices—each optimized toward solving only a particular class of problems. The dream of a universal quantum computer and the eventual vision for a fully functional quantum robot remain unrealized today. Several proposals for quantum robot architectures have previously been suggested [Benioff, 1998a,b, Dong et al., 2006], which describe the required hardware components and potential functionalities of future quantum robots.

What's certain is that an overall new generation of technologies will need to be developed for quantum robots to be realized—quantum software, quantum models of computation, quantum programming languages, and even quantum instruction set architectures.

In a more universal scheme of quantum computing, general-purpose quantum processors would entail the specification of a quantum instruction set architecture (qISA) [Smith et al., 2016]. This qISA would form the interface between quantum software and quantum hardware. For quantum software, it would specify the assembly instructions for compilation of code written in high-level quantum programming languages. For quantum hardware, it would specify the arithmetic and logic operations, as well as memory access operations that quantum micro-architectural implementations of the qISA would be required to support. Additionally, a sufficiently holistic and sound design methodology for application-specific synthesized quantum hardware could emerge. One possibility is a quantum variant of the design methodology for application-specific instruction set processors (ASIP) [Glökler and Meyr, 2004].

Once workable quantum implementations are attained, optimizing the hardware for performance, resource usage, and energy consumption will be useful. One way to achieve hardware optimization could be through quantum logic synthesis [Banerjee, 2010, Hayes and Markov, 2006, Hung et al., 2006, Shende et al., 2006]. The runtime and memory usage of the software could be further optimized via implementation of software algorithms discussed previously in this book. The full list of components necessary before the hardware design and implementation of quantum computers and quantum robots can reach their full potential remains to be determined.

7.7 CHAPTER SUMMARY

This chapter introduced the DiVincenzo Definition and Mosca Classification as frameworks one can use to gauge the progress of quantum implementations over time. Several implementation strategies were discussed, and the D-Wave Machine was evaluated in greater detail. The D-Wave discussion is a case study that illustrates how future quantum implementations can be evaluated as time progresses. Frameworks like DiVincenzo and Mosca can be used to gauge the field of quantum computing, and its potential progress toward the future implementation of quantum robotics. Specific component requirements necessary for the development of quantum robotics were also discussed.

Chapter Key Points

- The DiVincenzo definition identifies the physical requirements for the quantum circuit model of computation to be implemented.

- The Mosca classification is comprised of five increasingly restrictive categories for determining whether a machine can be classified as a quantum computer.

- To date, no quantum implementation has met all the qualifications under either the DiVincenzo definition or the Level-5 Mosca classification of a quantum computer.

- A wide variety of physical models have been proposed for quantum computation, harnessing a range of atomic or subatomic particles as qubits.

- The D-Wave set of machines have been validated by many research groups as quantum computers that rely on intrinsically quantum effects to perform calculations, though there is still ongoing debate about fair ways to test them.

- The D-Wave (and other commercial or academic implementations) are not yet able to harness the full potential computational power of quantum mechanics.

- A quantum robot powered by a universal quantum computer would require components such as quantum software, quantum models of computation, quantum programming languages, and quantum instruction set architectures.

CHAPTER 8

Conclusion

Our work has broadly explored the applications of quantum mechanisms to robotics, consolidating the previously fragmented discussions across disciplines, each of which contributes a piece to the emerging field of quantum robotics. While much of our exploration of possibilities has been theoretical (as implementations of relevant quantum engineering are still far from practical), we have suggested some of the key ways that developments in quantum science and engineering will impact the world of robotics, much as they have already begun impacting other fields.

Our journey into quantum robotics began with fundamental quantum mechanics principles, exploring the qubit representation of an atom and the representation's capabilities to develop potentially breakthrough hardware possibilities for robotics. Advances in the underlying hardware storage and power systems of robots could have a tremendous impact on functionality, as these have classically been some of the bottlenecks of embedded systems design. Qubit representation of data can hypothetically be more memory and energy efficient than classical computer binary bits. By storing data within quantum superposition, the quantum memory may scale to represent exponentially more data in the same number of bits than classically possible. By manipulating bits approaching the Landauer Limit of energy, quantum circuits can theoretically use millions of times less energy than classical computer circuits.

If the quantum hardware upgrades for robots sound useful, the quantum software upgrades inspire even more cause for excitement. Quantum parallelism, Grover's algorithm, and adiabatic optimization are key approaches to providing asymptotic speedup for simple, naïve brute force search in a parameter space—a speedup in one of the most fundamental computer science problems. These methods offer quadratically better runtimes for a multitude of classical robotic algorithms that rely on search as a primary aspect of their design. When extended to robotic planning algorithms, Grover Tree Search for both the uninformed and informed cases similarly exhibits possible speedups over classical approaches.

In addition to enhancements in search and planning, quantum mechanisms may provide a boost to robot learning. Quantum robotic agents can potentially learn quadratically faster than classical agents using a hypothetical "quantum active learning" algorithm. This enhancement in robot learning could allow robots to comprehend and perform well in more complex and unstructured environments than previously possible. Quantum robots may not be able to query their environment in superposition, but they are able to compute over large data sets of environment percepts in fewer iterations than a classical robot and thereby reason more robustly over its uncertainty factors.

Our broader survey of quantum machine learning also illustrated the many speedups a significant proportion of machine learning algorithms are expected to experience when deployed in quantum media. Being able to learn from data faster will allow for more sophisticated perception algorithms in which a quantum robot could more quickly learn complex models of an environment. Given the complexity of many modeling problems in robotics (such as that of manipulation and contact force dynamics), the enhancements of quantum machine learning will be welcome. Robots will likely be able to perceive patterns in data faster and more robustly than previously possible, a boon for more precise sensing and control.

In addition, many machine learning algorithms become more general in the quantum world, not requiring assumptions such as convex loss functions in SVMs or approximations such as contrastive divergence in Deep Boltzmann Machines. The generality allows for more modeling power of algorithms and added capability to learn more complex real-world functions than classical robots. A deeper, more intricate representation of an environment would allow robots to function better in a world governed by immense complexity.

Models discussed in this book also shed light on an interesting new possibility: robots operating in quantum environments or controlling quantum phenomena. qMDPs and QOMDPs extend the classical capabilities of MDPs and POMDPs respectively, facilitating planning and control strategies for robots manipulating quantum environments. HQMMs extended HMMs to allow robust filtering of quantum percepts. In addition, our survey explored a variety of control models such as Bilinear Models and Markovian Master Equations that extend robot controls primitives to the quantum world.

The actual implementation of quantum robotic technology is very much in its infancy. After all, we have yet to fully develop a working quantum computer, let alone a quantum robot that uses one. However, several components of quantum robots have been prototyped. For example, different types of optical artificial neural networks have been considered, different adiabatic optimization approaches have been explored for various machine learning algorithms, and many proposals are on the table for implementing active learning algorithms, HQMMs, and other such frameworks. The commercial implementation for quantum computing which has made the most progress, D-Wave, was explored as a case study in our book, highlighting various frameworks for defining a quantum computer, and providing a benchmark for understanding how far the field has progressed and what implementation challenges remain for the field of quantum robotics to achieve its full potential.

This literature review is just the beginning for the community. Much work clearly remains to be done before one can construct a quantum robot. Also (as Aaronson [2015] notes), many of the algorithms and methods discussed in this survey have been shown to work only in mathematical theory, and, even then, are limited in their scope and application. They cannot yet be widely applied to the point that, as some newspapers erroneously suggest, quantum computers will be able to replace all classical ones overnight. Thus it is imperative that one understands the "fine print" carefully articulated in the research papers which comprise the field to understand the mathematical preconditions necessary to be met before a particular algorithm or method can be applied.

While still in its infancy, the field of quantum robotics still holds great promise. A quantum robot is likely to possess interesting properties, some of which we've detailed in this book. As we better understand the underlying components needed to build a quantum robot, our understanding of the properties of a quantum robot will increase as well. Current theory suggests that, compared with the robots of today, quantum robotics will be able to interact with more complex environments, operate with greater energy and memory efficiently, learn faster and more robustly, and have the ability to manipulate and control quantum phenomena. Such robots could potentially contribute to productivity in the workplace, assist in the development of scientific research, and improve our quality of life. The emerging world of quantum robotics promises to be an exciting one. We hope our work has inspired you to delve deeper into the academic literature and learn more about it.

APPENDIX A

Cheatsheet of Quantum Concepts Discussed

	Quantum Concept	Potential Impact on Robotics
Chapter 2: Background on Quantum Mechanics Quantum Operating Principles (QOPs) lay the foundation for discussion of Quantum Robotics	Quantum Superposition	Probabilistically represents more data than classical bits
	Quantum Fast Fourier Transform (QFFT)	Requires fewer circuit elements compared to classical FFT
	Hadamard Transform	Building block for quantum parallelism
	Grover's Search Algorithm	Quadratic speedup in search problems
	Quadratic Unconstrained Binary Optimization (QUBO)	Potentially enables faster search and optimization for robotic computational problems
	Shor's Algorithm	Impacts public-key cryptography by breaking RSA encryption
	Quantum Teleportation	Transports qubit information with less effort than today
Chapter 3: Quantum Search Faster search capabilities	Grover Tree Search	Speedup over classical tree search planning, powered by Grover's Algorithm
	Quantum STRIPS	Solves STRIPS-encoded problem via quantum adiabatic hardware
Chapter 4: Quantum Agent Models Decision making and active learning	qMDPs and QOMDPs	Enables robot to plan in environments governed by quantum dynamics
	Quantum Active Learning	Enables robots to learn faster in complex but structured classical environments
	Single Photon Decision Maker	Potential speedup for multi-armed bandit problems using photonic hardware

	Quantum Concept	Potential Impact on Robotics
Chapter 5: Machine Learning Mechanisms for Quantum Robotics Improved speed and performance in machine learning to process large, complex datasets	Quantum Memory	More efficient storage capabilities
	Lloyd's Algorithm	Faster computation of inner product between two vectors
	Generalized Quantum Dot Product	Approximates arbitrary distance functions
	Hamiltonian Simulation	Casts machine learning algorithms for simulation on quantum computers
	Quantum Principal Component Analysis	Exponential speedup over classical version
	HLL Algorithm	Efficiently analyze linear systems for quantum regression
	Quantum Clustering	Speedup enabling robot to understand more complex percepts
	Quantum Support Vector Machine (SVM)	Scale to much larger datasets Increases classification accuracy by finding more robust decision boundaries
	Quantum Bayesian Networks	Richer representation of robotic interaction by scaling to larger graphical structures
	Quantum Associative Memories	Additional storage capacity compared to Hopfield Networks
	Quantum Perceptron	Incorporates nonlinearity via quantum measurement or Feynman path integrals
	Quantum Deep Learning	Enables more efficient training of Boltzmann Machines via quantum amplitude estimation
	Quantum Manifold Learning	Speedup for applications in robot navigation, dynamic visual tracking and motion synthesis
	Quantum Boosting (QBoost)	Generalizes to avoid requirement of convexity present in classical Adaboost

	Quantum Concept	Potential Impact on Robotics
Chapter 6: Quantum Filtering and Control Classical filtering and control algorithms extended to handle quantum environments	Projective Measurement (i.e., von Neumann Measurement)	Extracts instantaneous measurement from quantum system
	Continuous Measurement	Extracts continuous measurement from quantum system via an ancilla system
	Hidden Quantum Markov Models (HQMMs)	Filters system governed by quantum dynamics
	Quantum Kalman Filter	Allows indirect information acquisition of a quantum system via non-demolition measurements
	Bilinear Model (BLM)	Describes closed quantum control systems including molecular systems and spin systems
	Markovian Master Equation (MME)	Describes state of both closed and open quantum system with applications in quantum error correction and spin squeezing
	Stochastic Master Equation (SME)	Allows incorporation of feedback into quantum control systems
	Linear Quantum Stochastic Differential Equation (LQSDE)	Control of systems using Heisenberg Model
Chapter 7: Current Strategies for Quantum Implementation Discussion of progress in quantum implementations	DiVincenzo Definition	Circuit-based definition for a universal quantum computer Currently no known implementation fulfills this definition
	Mosca Classification	Five increasingly restrictive definitions for classifying a quantum computer. Currently the most advanced quantum computers are Level 3 Mosca machines

Bibliography

Scott Aaronson. Read the fine print. *Nature Physics*, 11(4), pages 291–293, 2015. DOI: 10.1038/nphys3272. 100

Steven Adachi and Maxwell Henderson. Application of quantum annealing to training of deep neural networks, 2015. 67

Charlene Ahn, Andrew C. Doherty, and Andrew J. Landahl. Continuous quantum error correction via quantum feedback control. *Physical Review A*, 65(4), page 042301, 2002. DOI: 10.1103/physreva.65.042301. 86

Esma Aïmeur, Gilles Brassard, and Sébastien Gambs. Quantum speed-up for unsupervised learning. *Machine Learning*, 90(2), pages 261–287, 2013. DOI: 10.1007/s10994-012-5316-5. 55, 69

Tameem Albash, Walter Vinci, Anurag Mishra, Paul A. Warburton, and Daniel A. Lidar. Consistency tests of classical and quantum models for a quantum annealer. *Physical Review A*, 91(4), 2015. DOI: 10.1103/PhysRevA.91.042314. 95

Jacopo Aleotti and Stefano Caselli. Trajectory clustering and stochastic approximation for robot programming by demonstration. In *Intelligent Robots and Systems, (IROS). IEEE/RSJ International Conference on*, pages 1029–1034, 2005. DOI: 10.1109/iros.2005.1545365. 56

Domenico D. Alessandro and Mohammed Dahleh. Optimal control of two-level quantum systems. *Automatic Control, IEEE Transactions on*, 46(6), pages 866–876, 2001. DOI: 10.1109/9.928587. 85

Armen E. Allahverdyan and Th. M. Nieuwenhuizen. Breakdown of the landauer bound for information erasure in the quantum regime. *Physical Review E*, 64(5), 2001. DOI: 10.1103/physreve.64.056117.

Michael Altaisky. Quantum neural network, 2001. 93

Mikhail V. Altaisky, Nadezhda N. Zolnikova, Natalia E. Kaputkina, Victor A. Krylov, Yurii E. Lozovik, and Nikesh S. Dattani. Towards a feasible implementation of quantum neural networks using quantum dots, 2015. DOI: 10.1063/1.4943622. 93

Janet Anders, Saroosh Shabbir, Stefanie Hilt, and Eric Lutz. Landauer's principle in the quantum domain. *Electronic Proceedings in Theoretical Computer Science*, 26, pages 13–18, 2010. DOI: 10.4204/eptcs.26.2.

Brian D.O. Anderson. The realization problem for hidden Markov models. *Mathematics of Control, Signals and Systems*, 12(1), pages 80–120, 1999. DOI: 10.1007/pl00009846. 79

Davide Anguita, Sandro Ridella, Fabio Rivieccio, and Rodolfo Zunino. Quantum optimization for training support vector machines. *Neural Networks*, 16(5), pages 763–770, 2003. DOI: 10.1016/s0893-6080(03)00087-x. 58

Alán Aspuru-Guzik and Philip Walther. Photonic quantum simulators. *Nature Physics*, 8(4), pages 285–291, 2012. DOI: 10.1038/nphys2253. 43

James Andrew Bagnell, David Bradley, David Silver, Boris Sofman, and Anthony Stentz. Learning for autonomous navigation. *Robotics and Automation Magazine, IEEE*, 17(2), pages 74–84, 2010. DOI: 10.1109/mra.2010.936946. 47

Christel Baier and Joost-Pieter Katoen. *Principles of Model Checking (Representation and Mind Series)*. The MIT Press, 2008. 37

Anindita Banerjee. *Synthesis, Optimization and Testing of Reversible and Quantum Circuits*. Ph.D. thesis, Jaypee Institute of Information Technology, Noida, India, 2010. http://hdl.handle.net/10603/2425 97

Jennifer Barry, Daniel T. Barry, and Scott Aaronson. Quantum partially observable Markov decision processes. *Physical Review A*, 90(3), page 032311, 2014. DOI: 10.1103/physreva.90.032311. 33, 36, 38, 39

Elizabeth Behrman, John Niemel, James Steck, and Steven R. Skinner. A quantum dot neural network. In *Proc. of the 4th Workshop on Physics of Computation*, pages 22–24, Citeseer, 1996. 66, 93

Viacheslav Belavkin. Measurement, filtering and control in quantum open dynamical systems. *Reports on Mathematical Physics*, 43(3), pages A405–A425, 1999. DOI: 10.1016/s0034-4877(00)86386-7.

Mikhail Belkin and Partha Niyogi. Laplacian eigenmaps and spectral techniques for embedding and clustering. In *Advances in Neural Information Processing Systems 14*, pages 585–591, MIT Press, 2001. 69

Richard Bellman. A markovian decision process. Technical report, DTIC Document, 1957. DOI: 10.1512/iumj.1957.6.56038. 33, 34

Marcello Benedetti, John Realpe-Gómez, Rupak Biswas, and Alejandro Perdomo-Ortiz. Estimation of effective temperatures in a quantum annealer and its impact in sampling applications: A case study towards deep learning applications, 2015. DOI: 10.1103/physreva.94.022308. 67

Yoshua Bengio. Learning deep architectures for ai. *Foundations and Trends® in Machine Learning*, 2(1), pages 1–127, 2009. DOI: 10.1561/2200000006. 47, 64, 66

Yoshua Bengio, Yann LeCun, et al. Scaling learning algorithms towards ai. *Large-scale Kernel Machines*, 34(5), 2007. 64

Paul Benioff. Quantum robots and environments. *Physical Review A*, 58(2), pages 893–904, 1998a. DOI: https://doi.org/10.1103/PhysRevA.58.893. 96

Paul Benioff. Quantum robots plus environments. In *Proc. of the 4th International Conference on Quantum Communication, Computing, and Measurement (QCM'92)*, volume 2, pages 3–9, Evanston, IL, August 22–27 1998b. Kluwer Academic Publishers. DOI: 10.1007/0-306-47097-7_1. 96

Charles H. Bennett. Logical reversibility of computation. *IBM Journal of Research and Development*, 17(6), pages 525–532, 1973. DOI: 10.1147/rd.176.0525. 10

Dominic W. Berry, Graeme Ahokas, Richard Cleve, and Barry C. Sanders. Efficient quantum algorithms for simulating sparse hamiltonians. *Communications in Mathematical Physics*, 270(2), pages 359–371, 2007. DOI: 10.1007/s00220-006-0150-x. 50

Antoine Bérut, Artak Arakelyan, Artyom Petrosyan, Sergio Ciliberto, Raoul Dillenschneider, and Eric Lutz. Experimental verification of landauer/'s principle linking information and thermodynamics. *Nature*, 483(7388), pages 187–189, 2012. DOI: 10.1038/nature10872. 10

Alessandro Bisio, Giulio Chiribella, Giacomo Mauro D'Ariano, Stefano Facchini, and Paolo Perinotti. Optimal quantum learning of a unitary transformation. *Physical Review A*, 81(3), pages 032324, 2010. DOI: 10.1103/physreva.81.032324.

Joydeep Biswas and Manuela Veloso. Wifi localization and navigation for autonomous indoor mobile robots. In *Robotics and Automation (ICRA), IEEE International Conference on*, pages 4379–4384, 2010. DOI: 10.1109/robot.2010.5509842. 47

Rainer Blatt and Christian F. Roos. Quantum simulations with trapped ions. *Nature Physics*, 8(4), pages 277–284, 2012. DOI: 10.1038/nphys2252. 43

Sergio Boixo, Tameem Albash, Federico M. Spedalieri, Nicholas Chancellor, and Daniel A. Lidar. Experimental signature of programmable quantum annealing. *Nature Communications*, 4, 2013. DOI: 10.1038/ncomms3067. 94

Sergio Boixo, Troels F. Rønnow, Sergei V. Isakov, Zhihui Wang, David Wecker, Daniel A. Lidar, John M. Martinis, and Matthias Troyer. Evidence for quantum annealing with more than one hundred qubits. *Nature Physics*, 10(3), pages 218–224, 2014. DOI: 10.1038/nphys2900. 94

Max Born and Vladimir Fock. Beweis des adiabatensatzes. *Zeitschrift für Physik*, 51(3–4), pages 165–180, 1928. ISSN 0044-3328. DOI: 10.1007/BF01343193. 17

Léon Bottou. Online algorithms and stochastic approximations. In David Saad, Ed., *Online Learning and Neural Networks*. Cambridge University Press, Cambridge, UK, 1998. `http://leon.bottou.org/papers/bottou-98x` revised, oct 2012. DOI: 10.1017/cbo9780511569920.003. 33

Michel Boyer, Gilles Brassard, Peter Høyer, and Alain Tapp. Tight bounds on quantum searching. In *Quantum Computing*, pages 187–199. Wiley-Blackwell, 2004. DOI: 10.1002/3527603093.ch10. 16

Darius Braziunas. Pomdp solution methods. *University of Toronto, Tech. Rep*, 2003. 35

Hans J Briegel and Gemma De las Cuevas. Projective simulation for artificial intelligence. *Scientific Reports*, 2, 2012. DOI: 10.1038/srep00400. 40

Hans J. Briegel, Tommaso Calarco, Dieter Jaksch, Juan Ignacio Cirac, and Peter Zoller. Quantum computing with neutral atoms. *Journal of Modern Optics*, 47(2–3), pages 451–451, 2000. DOI: 10.1080/09500340008244052. 92

Emma Brunskill and Nicholas Roy. Slam using incremental probabilistic pca and dimensionality reduction. In *Robotics and Automation (ICRA). Proc. of the 2005 IEEE International Conference on*, pages 342–347, 2005. DOI: 10.1109/robot.2005.1570142. 51

Paul I. Bunyk, Emile M. Hoskinson, Mark W. Johnson, Elena Tolkacheva, Fabio Altomare, Andrew J. Berkley, Roy Harris, J. P. Hilton, T. Lanting, and J. Whittaker. Architectural considerations in the design of a superconducting quantum annealing processor. *IEEE Transactions on Applied Superconductivity*, 24(4), pages 1–10, 2014. DOI: 10.1109/TASC.2014.2318294. 93

Tom Bylander. The computational complexity of propositional strips planning. *Artificial Intelligence*, 69(1), pages 165–204, 1994. DOI: 10.1016/0004-3702(94)90081-7. 28

Huaixin Cao, Feilong Cao, and Dianhui Wang. Quantum artificial neural networks with applications. *Information Sciences*, 290, pages 1–6, 2015. DOI: 10.1016/j.ins.2014.08.033.

Anthony R. Cassandra. A survey of pomdp applications. In *Working Notes of AAAI 1998 Fall Symposium on Planning with Partially Observable Markov Decision Processes*, volume 1724, Citeseer, 1998. 33

Lawrence Cayton. Algorithms for manifold learning. *Univ. of California at San Diego Tech. Rep.*, pages 1–17, 2005. 69

Giulio Chiribella, Giacomo Mauro D'ariano, and Massimiliano F. Sacchi. Optimal estimation of group transformations using entanglement. *Physical Review A*, 72(4), pages 042338, 2005. DOI: 10.1103/physreva.72.042338.

Roberto Cipolla, Sebastiano Battiato, and Giovanni Maria Farinella. *Machine Learning for Computer Vision*. Springer, 2013. DOI: 10.1007/978-3-642-28661-2. 47

Juan Ignacio Cirac and Peter Zoller. A scalable quantum computer with ions in an array of microtraps. *Nature*, 404(6778), pages 579–581, 2000. DOI: 10.1038/35007021. 92

Lewis A. Clark, Wei Huang, Thomas M. Barlow, and Almut Beige. Hidden quantum Markov models and open quantum systems with instantaneous feedback. In *ISCS 2014: Interdisciplinary Symposium on Complex Systems*, pages 143–151, Springer, 2015. DOI: 10.1007/978-3-319-10759-2_16. 78

Jorge Cortes, Léonard Jaillet, and Thierry Siméon. Disassembly path planning for complex articulated objects. *Robotics, IEEE Transactions on*, 24(2), pages 475–481, 2008. DOI: 10.1109/tro.2008.915464. 33

A. D. Córcoles, Easwar Magesan, Srikanth J. Srinivasan, Andrew W. Cross, M. Steffen, Jay M. Gambetta, and Jerry M. Chow. Demonstration of a quantum error detection code using a square lattice of four superconducting qubits. *Nature Communications*, 6, 2015. DOI: 10.1038/ncomms7979. 93

Nikesh S. Dattani and Nathaniel Bryans. Quantum factorization of 56153 with only 4 qubits, 2014. 20

Erick Delage, Honglak Lee, and Andrew Y. Ng. A dynamic Bayesian network model for autonomous 3d reconstruction from a single indoor image. In *Computer Vision and Pattern Recognition, IEEE Computer Society Conference on*, volume 2, pages 2418–2428, 2006. DOI: 10.1109/cvpr.2006.23. 60, 62

Vasil S. Denchev, Sergio Boixo, Sergei V. Isakov, Nan Ding, Ryan Babbush, Vadim Smelyanskiy, John Martinis, and Hartmut Neven. What is the computational value of finite range tunneling? 2015. DOI: 10.1103/physrevx.6.031015. 96

Misha Denil and Nando De Freitas. Toward the implementation of a quantum rbm. In *NIPS Deep Learning and Unsupervised Feature Learning Workshop*, 2011. 67

David Deutsch. Uncertainty in quantum measurements. *Physical Review Letters*, 50(9), page 631, 1983. DOI: 10.1103/physrevlett.50.631. 75

David Deutsch and Richard Jozsa. Rapid solution of problems by quantum computation. In *Proc. of the Royal Society of London A: Mathematical, Physical and Engineering Sciences*, volume 439, pages 553–558, The Royal Society, 1992. DOI: 10.1098/rspa.1992.0167. 14

Thomas G. Dietterich. Machine-learning research. *AI magazine*, 18(4), page 97, 1997. 47

David DiVincenzo. The physical implementation of quantum computation. *Fortschr. Phys.*, 48, pages 771–783, 2000. DOI: 10.1002/1521-3978(200009)48:9/11<771::AID-PROP771>3.0.CO;2-E. 89

Daoyi Dong, Chunlin Chen, Chenbin Zhang, and Zonghai Chen. Quantum robot: Structure, algorithms and applications. *Robotica*, 24(4), pages 513–521, July 2006. DOI: 10.1017/S0263574705002596. 96

Dianbiao Dong and Ian R. Petersen. Quantum control theory and applications: A survey. *Control Theory and Applications, IET*, 4(12), pages 2651–2671, 2010. DOI: 10.1049/iet-cta.2009.0508. 83

Arnaud Doucet, Nando de Freitas, Kevin P. Murphy, and Stuart J. Russell. Rao-blackwellised particle filtering for dynamic Bayesian networks. In *Proc. of the 16th Conference on Uncertainty in Artificial Intelligence*, UAI'00, pages 176–183, San Francisco, CA, Morgan Kaufmann Publishers Inc., 2000. http://dl.acm.org/citation.cfm?id=647234.720075 DOI: 10.1007/978-1-4757-3437-9_24. 60, 62

Lu-Ming Duan and Guang-Can Guo. Probabilistic cloning and identification of linearly independent quantum states. *Physical Review Letters*, 80(22), page 4999, 1998. DOI: 10.1103/physrevlett.80.4999. 65

Pierre Duhamel and Martin Vetterli. Fast fourier transforms: A tutorial review and a state of the art. *Signal Processing*, 19(4), pages 259–299, 1990. http://www.sciencedirect.com/science/article/pii/016516849090158U DOI: doi:10.1016/0165-1684(90)90158-U.

Vincent Dumoulin, Ian J. Goodfellow, Aaron Courville, and Yoshua Bengio. On the challenges of physical implementations of rbms, 2013. 67

Christoph Dürr, Mark Heiligman, Peter Høyer, and Mehdi Mhalla. Quantum query complexity of some graph problems. In *Automata, Languages and Programming*, pages 481–493. Springer, 2004. DOI: 10.1007/978-3-540-27836-8_42. 69

Aaron Edsinger-Gonzales and Jeff Weber. Domo: A force sensing humanoid robot for manipulation research. In *Humanoid Robots, 4th IEEE/RAS International Conference on*, volume 1, pages 273–291, 2004. DOI: 10.1109/ichr.2004.1442127. 47

Alexandr A. Ezhov and Dan Ventura. Quantum neural networks. In *Future Directions for Intelligent Systems and Information Sciences*, pages 213–235, Springer, 2000. DOI: 10.1007/978-3-7908-1856-7_11. 67

Jean Faber and Gilson A. Giraldi. Quantum models for artificial neural networks. *Electronically*, 2002. http://arquivosweb.lncc.br/pdfs/QNN-Review.pdf 66, 67

Richard E. Fikes and Nils J. Nilsson. Strips: A new approach to the application of theorem proving to problem solving. *Artificial Intelligence*, 2(3-4), pages 189–208, 1971. DOI: 10.1016/0004-3702(71)90010-5. 28

Maria Fox, Malik Ghallab, Guillaume Infantes, and Derek Long. Robot introspection through learned hidden Markov models. *Artificial Intelligence*, 170(2), pages 59–113, 2006. DOI: 10.1016/j.artint.2005.05.007. 78

Yoav Freund, Robert Schapire, and Naoke Abe. A short introduction to boosting. *Journal-Japanese Society For Artificial Intelligence*, 14(771–780), page 1612, 1999. 70

John Gantz and David Reinsel. The digital universe in 2020: Big data, bigger digital shadows, and biggest growth in the far east. *IDC iView: IDC Analyze the Future*, 2007, pages 1–16, 2012. 48

Chris Gaskett and Gordon Cheng. Online learning of a motor map for humanoid robot reaching. In *In Proc. of the 2nd International Conference on Computational Intelligence, Robotics and Autonomous Systems (CIRAS)*, 2003. 47

Nicolas Gisin. Quantum measurements and stochastic processes. *Physical Review Letters*, 52(19), page 1657, 1984. DOI: 10.1103/physrevlett.52.1657. 75

Ross Glashan, Kaijen Hsiao, Leslie Pack Kaelbling, and Tomás Lozano-Pérez. Grasping POMDPs: Theory and experiments. In *Robotics Science and Systems Manipulation Workshop: Sensing and Adapting to the Real World*, 2007. http://www.willowgarage.com/sites/def ault/files/paper__grasping_pomdps_theory_and_experiments__glashan.pdf 33

Tilman Glökler and Heinrich Meyr. *Design of Energy-Efficient Application-Specific Instruction Set Processors*. Kluwer Academic Publishers, Dordrecht, The Netherlands, 2004. DOI: 10.1007/b105292. 97

Raia Hadsell, Pierre Sermanet, Jan Ben, A. Erkan, Jeff Han, Beat Flepp, Urs Muller, and Yann LeCun. Online learning for offroad robots: Using spatial label propagation to learn long-range traversability. In *Proc. of Robotics: Science and Systems (RSS)*, volume 11, page 32, 2007. DOI: 10.15607/rss.2007.iii.003. 47

Hani Hagras. Online learning techniques for prototyping outdoor mobile robots operating in unstructured environments, 2001. 47

Aram W. Harrow, Avinatan Hassidim, and Seth Lloyd. Quantum algorithm for linear systems of equations. *Physical Review Letters*, 103(15), page 150502, 2009. DOI: 10.1103/physrevlett.103.150502. 49, 52

Ioannis Havoutis and Subramanian Ramamoorthy. Motion synthesis through randomized exploration on submanifolds of configuration space. In *RoboCup 2009: Robot Soccer World Cup XIII*, pages 92–103, Springer, 2010. DOI: 10.1007/978-3-642-11876-0_9. 70

John P. Hayes and Igor L. Markov. Quantum approaches to logic circuit synthesis and testing. Technical Report ADA454812, Defense Technical Information Center, Fort Belvoir, VA, 2006. `http://oai.dtic.mil/oai/oai?verb=getRecord&metadataPrefix=html&identifier=ADA454812` 97

Itay Hen, Joshua Job, Tameem Albash, Troels F. Rønnow, Matthias Troyer, and Daniel Lidar. Probing for quantum speedup in spin glass problems with planted solutions, 2015. DOI: 10.1103/PhysRevA.92.042325. 96

Geoffrey E. Hinton. Training products of experts by minimizing contrastive divergence. *Neural Computation*, 14(8), pages 1771–1800, 2002. DOI: 10.1162/089976602760128018. 66

Geoffrey E. Hinton and Terrence J. Sejnowski. Learning and relearning in boltzmann machines. In David E. Rumelhart, James L. McClelland, and CORPORATE PDP Research Group, Eds., *Parallel Distributed Processing: Explorations in the Microstructure of Cognition, Vol. 1*, pages 282–317, MIT Press, Cambridge, MA, 1986. `http://dl.acm.org/citation.cfm?id=104279.104291` 63

Geoffrey E. Hinton, Simon Osindero, and Yee-Whye Teh. A fast learning algorithm for deep belief nets. *Neural Computation*, 18(7), pages 1527–1554, 2006. DOI: 10.1162/neco.2006.18.7.1527. 64

Kazuo Hiraki. Machine learning in human-robot interaction. *Advances in Human Factors/Ergonomics*, 20, pages 1127–1132, 1995. DOI: 10.1016/s0921-2647(06)80178-4. 47

Jeongmin Hong, Brian Lambson, Scott Dhuey, and Jeffrey Bokor. Experimental test of landauer's principle in single-bit operations on nanomagnetic memory bits. *Science Advances*, 2(3), 2016. `http://advances.sciencemag.org/content/2/3/e1501492` DOI: 10.1126/sciadv.1501492. 10

John J. Hopfield. Neural networks and physical systems with emergent collective computational abilities. *Proc. of the National Academy of Sciences*, 79(8), pages 2554–2558, 1982. DOI: 10.1073/pnas.79.8.2554. 63

Mark Horowitz and Joel Burdick. Interactive non-prehensile manipulation for grasping via pomdps. In *Robotics and Automation (ICRA), International Conference on*, pages 3257–3264, IEEE, 2013. DOI: 10.1109/icra.2013.6631031. 33

Geir E. Hovland, Pavan Sikka, and Brenan J. McCarragher. Skill acquisition from human demonstration using a hidden Markov model. In *Robotics and Automation,*

Proc. of the International Conference on, volume 3, pages 2706–2711, IEEE, 1996. DOI: 10.1109/robot.1996.506571. 78

Kaijen Hsiao, Leslie Pack Kaelbling, and Tomas Lozano-Perez. Grasping pomdps. In *Robotics and Automation, International Conference on*, pages 4685–4692, IEEE, 2007. DOI: 10.1109/robot.2007.364201. 33

David Hsu, Wee Sun Lee, and Nan Rong. A point-based pomdp planner for target tracking. In *Robotics and Automation, (ICRA) International Conference on*, pages 2644–2650, IEEE, 2008. DOI: 10.1109/robot.2008.4543611. 33

Richard Hughes, Gary D. Doolen, David Awschalom, Carlton M. Caves, Michael Chapman, Robert Clark, David G. Cory, David DiVincenzo, Artur Ekert, P. Chris Hammel, and Paul G. Kwiat. A quantum information science and technology roadmap part 1: Quantum computation, 2004. 92

William N. N. Hung, Xiaoyu Song, Guowu Yang, Jin Yang, and Marek Perkowski. Optimal synthesis of multiple output boolean functions using a set of quantum gates by symbolic reachability analysis. *IEEE Transactions on Computer-Aided Design of Integrated Circuits and Systems*, 25(9), pages 1652–1663, 2006. DOI: 10.1109/TCAD.2005.858352. 97

Kurt Jacobs and Daniel A. Steck. A straightforward introduction to continuous quantum measurement. *Contemporary Physics*, 47(5), pages 279–303, 2006. DOI: 10.1080/00107510601101934. 76

Matthew R. James, Hendra I. Nurdin, and Ian R. Petersen. H-infinity control of linear quantum stochastic systems, 2007. 87

Wei Ji, Dean Zhao, Fengyi Cheng, Bo Xu, Ying Zhang, and Jinjing Wang. Automatic recognition vision system guided for apple harvesting robot. *Computers and Electrical Engineering*, 38(5), pages 1186–1195, 2012. DOI: 10.1016/j.compeleceng.2011.11.005. 56, 59

Mark W. Johnson, Mohammad H. S. Amin, Suzanne Gildert, Trevor Lanting, Firas Hamze, Neil Dickson, R. Harris, Andrew J. Berkley, Jan Johansson, Paul Bunyk, Erin M. Chapple, C. Enderud, Jeremy P. Hilton, Kamran Karimi, Eric Ladizinsky, Nicholas Ladizinsky, Travis Oh, Ilya Perminov, Chris Rich, Murray C. Thom, E. Tolkacheva, Colin J. S. Truncik, Sergey Uchaikin, Jun Wang, B. Wilson, and Geordie Rose. Quantum annealing with manufactured spins. *Nature*, 473(7346), pages 194–198, 2011. DOI: 10.1038/nature10012. 94

Ari Juels and Bonnie Wong. The interplay of neuroscience and cryptography: Technical perspective. *Communications of the ACM*, 57(5), pages 109–109, 2014. DOI: 10.1145/2594446. 48

Leslie Pack Kaelbling, Michael L. Littman, and Andrew W. Moore. Reinforcement learning: A survey. *Journal of Artificial Intelligence Research*, 4(1), pages 237–285, 1996. ISSN 1076-9757. http://dl.acm.org/citation.cfm?id=1622737.1622748 47

Leslie Pack Kaelbling, Michael L. Littman, and Anthony R. Cassandra. Planning and acting in partially observable stochastic domains. *Artificial Intelligence*, 101(1), pages 99–134, 1998. DOI: 10.1016/s0004-3702(98)00023-x. 33, 35

Rudolph Emil Kalman. A new approach to linear filtering and prediction problems. *Journal of Basic Engineering*, 82(1), pages 35–45, 1960. DOI: 10.1115/1.3662552. 80

Henry Kautz and Bart Selman. Satplan04: Planning as satisfiability. *Working Notes on the 5th International Planning Competition (IPC)*, pages 45–46, 2006. 30

Phillip Kaye, Raymond Laflamme, and Michele Mosca. An introduction to quantum computing, 2006. 13

Narongdech Keeratipranon, Frederic Maire, and Henry Huang. Manifold learning for robot navigation. *International Journal of Neural Systems*, 16(05), pages 383–392, 2006. DOI: 10.1142/s0129065706000780. 70

Richard Kelley, Alireza Tavakkoli, Christopher King, Monica Nicolescu, Mircea Nicolescu, and George Bebis. Understanding human intentions via hidden Markov models in autonomous mobile robots. In *Proc. of the 3rd ACM/IEEE International Conference on Human Robot Interaction*, pages 367–374, 2008. DOI: 10.1145/1349822.1349870. 78

David Kielpinski, Chris Monroe, and David J. Wineland. Architecture for a large-scale ion-trap quantum computer. *Nature*, 417(6890), pages 709–711, 2002. DOI: 10.1038/nature00784. 92

James King, Sheir Yarkoni, Mayssam M. Nevisi, Jeremy P. Hilton, and Catherine C. McGeoch. Benchmarking a quantum annealing processor with the time-to-target metric, 2015. 95

Sven Koenig and Reid G. Simmons. Xavier: A robot navigation architecture based on partially observable markov decision process models. In David Kortenkamp, R. Peter Bonasso, and Robin Murphy, Eds., *Artificial Intelligence and Mobile Robots*, pages 91–122, MIT Press, Cambridge, MA, 1998. http://dl.acm.org/citation.cfm?id=292092.292120 33

Pieter Kok, William J. Munro, Kae Nemoto, Timothy C. Ralph, Jonathan P. Dowling, and Gerard J. Milburn. Linear optical quantum computing with photonic qubits. *Reviews of Modern Physics*, 79(1), 2007. DOI: 10.1103/RevModPhys.79.135. 93

Daphne Koller and Nir Friedman. *Probabilistic Graphical Models: Principles and Techniques—Adaptive Computation and Machine Learning*. The MIT Press, 2009. 60

Thaddeus D. Ladd, Fedor Jelezko, Raymond Laflamme, Yasunobu Nakamura, Christopher Monroe, and Jeremy L. O'Brien. Quantum computers. *Nature*, 464(7285), pages 45–53, 2010. DOI: 10.1038/nature08812. 90

Rolf Landauer. Irreversibility and heat generation in the computing process. *IBM Journal of Research and Development*, 5(3), pages 183–191, 1961. DOI: 10.1147/rd.53.0183. 10

Trevor Lanting, A. J. Przybysz, Anatoly Y. Smirnov, Federico M. Spedalieri, Mohammad H. Amin, Andrew J. Berkley, R. Harris, Fabio Altomare, Sergio Boixo, Paul Bunyk, Neil Dickson, C. Enderud, Jeremy P. Hilton, Emile Hoskinson, Mark W. Johnson, Eric Ladizinsky, Nicholas Ladizinsky, Richard Neufeld, Travis Oh, Ilya Perminov, Chris Rich, Murray C. Thom, E. Tolkacheva, Sergey Uchaikin, A. B. Wilson, and Geordie Rose. Entanglement in a quantum annealing processor. *Physical Review X*, 4(2), 2014. DOI: 10.1103/PhysRevX.4.021041. 94

Marco Lanzagorta and Jeffrey Uhlmann. *Quantum Computer Science*. Synthesis Lectures on Quantum Computing. Morgan & Claypool Publishers, San Rafael, CA, 2008a. DOI: 10.2200/s00159ed1v01y200810qmc002.

Marco Lanzagorta and Jeffrey Uhlmann. Is quantum parallelism real? In *SPIE Defense and Security Symposium*, pages 69760W–69760W, International Society for Optics and Photonics, 2008b. DOI: 10.1117/12.778019. 15

Nicolas Le Roux and Yoshua Bengio. Representational power of restricted Boltzmann machines and deep belief networks. *Neural Computation*, 20(6), pages 1631–1649, 2008. DOI: 10.1162/neco.2008.04-07-510. 64

Yann LeCun, Koray Kavukcuoglu, and Clément Farabet. Convolutional networks and applications in vision. In *Circuits and Systems (ISCAS), Proc. of the International Symposium on*, pages 253–256, IEEE, 2010. DOI: 10.1109/iscas.2010.5537907. 47, 64, 67

Craig S. Lent, P. Douglas Tougaw, and Wolfgang Porod. Quantum cellular automata: The physics of computing with arrays of quantum dot molecules. *Physics and Computation, (PhysComp'94), Proc., Workshop on*, 1994. DOI: 10.1109/PHYCMP.1994.363705. 93

Ian Lenz, Honglak Lee, and Ashutosh Saxena. Deep learning for detecting robotic grasps. *The International Journal of Robotics Research*, 34(4-5), pages 705–724, 2015. DOI: 10.15607/rss.2013.ix.012. 47

Maciej Lewenstein. Quantum perceptrons. *Journal of Modern Optics*, 41(12), pages 2491–2501, 1994. DOI: 10.1080/09500349414552331. 65

Frank L. Lewis, Suresh Jagannathan, and A. Yesildirak. *Neural Network Control of Robot Manipulators and Nonlinear Systems*. Taylor & Francis, Inc., Bristol, PA, 1998. 62

Seth Lloyd, Masoud Mohseni, and Patrick Rebentrost. Quantum algorithms for supervised and unsupervised machine learning, 2013. 48, 49, 55

Seth Lloyd, Masoud Mohseni, and Patrick Rebentrost. Quantum principal component analysis. *Nature Physics*, 10(9), pages 631–633, 2014. DOI: 10.1038/nphys3029. 51, 53, 58

Suresh K. Lodha, Edward J. Kreps, David P. Helmbold, and Darren N. Fitzpatrick. Aerial lidar data classification using support vector machines (svm). In *3DPVT*, pages 567–574, 2006. DOI: 10.1109/3dpvt.2006.23. 47

Daniel Loss and David P. DiVincenzo. Quantum computation with quantum dots. *Physical Review A*, 57(1), 1998. DOI: 10.1103/PhysRevA.57.120. 93

Omid Madani, Steve Hanks, and Anne Condon. On the undecidability of probabilistic planning and related stochastic optimization problems. *Artificial Intelligence*, 147(1–2), pages 5–34, 2003. ISSN 0004-3702. http://www.sciencedirect.com/science/article/pii/S0004370202003788 Planning with uncertainty and incomplete information. DOI: 10.1016/S0004-3702(02)00378-8. 36

Frédéric Magniez, Ashwin Nayak, Jérémie Roland, and Miklos Santha. Search via quantum walk. *SIAM Journal on Computing*, 40(1), pages 142–164, 2011. DOI: 10.1137/090745854. 42, 43

Aditya Mahajan and Demosthenis Teneketzis. Multi-armed bandit problems. In *Foundations and Applications of Sensor Management*, pages 121–151, Springer, 2008. DOI: 10.1007/978-0-387-49819-5_6. 33, 44

Aleix M. Martínez and Jordi Vitria. Clustering in image space for place recognition and visual annotations for human-robot interaction. *Systems, Man, and Cybernetics, Part B: Cybernetics, IEEE Transactions on*, 31(5), pages 669–682, 2001. DOI: 10.1109/3477.956029. 56

Maja J. Matarić. *The Robotics Primer*. Intelligent Robotics and Autonomous Agents, MIT Press, 2007. https://books.google.com/books?id=WWJPjgz-jgEC 1

Drew McDermott, Malik Ghallab, Adele Howe, Craig Knoblock, Ashwin Ram, Manuela Veloso, Daniel Weld, and David Wilkins. Pddl-the planning domain definition language, 1998. 28

Catherine C. McGeoch. *Adiabatic Quantum Computation and Quantum Annealing: Theory and Practice*. Synthesis Lectures on Quantum Computing. Morgan & Claypool Publishers, 2012. DOI: 10.2200/S00585ED1V01Y201407QMC008. 96

Catherine C. McGeoch and Cong Wang. Experimental evaluation of an adiabatic quantum system for combinatorial optimization. *CF 13 Proc. of the ACM International Conference on Computing Frontiers*, 2013. DOI: 10.1145/2482767.2482797. 95

Alfred J. Menezes, Paul C. Van Oorschot, and Scott A. Vanstone. *Handbook of Applied Cryptography*. CRC press, 1996. `http://cacr.math.uwaterloo.ca/hac/` DOI: 10.1201/9781439821916. 20

Tammy Menneer and Ajit Narayanan. Quantum-inspired neural networks. *Department of Computer Science*, 329, page 1995, University of Exeter, Exeter, UK, Technical Report, 1995.

Tzvetan S. Metodi, Arvin I. Faruque, and Frederic T. Chong. *Quantum Computing for Computer Architects*, 2nd ed., Synthesis Lectures on Computer Architecture, Morgan & Claypool Publishers, San Rafael, CA, 2011. DOI: 10.2200/S00331ED1V01Y201101CAC013.

Marvin Minsky and Seymour A. Papert. Perceptrons: An essay in computational geometry, 1969. 63

Volodymyr Mnih, Koray Kavukcuoglu, David Silver, Andrei A. Rusu, Joel Veness, Marc G. Bellemare, Alex Graves, Martin Riedmiller, Andreas K. Fidjeland, Georg Ostrovski, et al. Human-level control through deep reinforcement learning. *Nature*, 518(7540), pages 529–533, 2015. DOI: 10.1038/nature14236. 67

Alex Monras, Almut Beige, and Karoline Wiesner. Hidden quantum Markov models and non-adaptive read-out of many-body states, 2010. 79

Alex Monras, Gerardo Adesso, Salvatore Marco Giampaolo, Giulia Gualdi, Gary B. Davies, and Fabrizio Illuminati. Entanglement quantification by local unitary operations. *Physical Review A*, 84(1), page 012301, 2011. DOI: 10.1103/physreva.84.012301. 79

Michele Mosca, Leo Mirani, and Gideon Lichfield. Why nobody can tell whether the world's biggest quantum computer is a quantum computer, 2014. `http://qz.com/194738/why-nobody-can-tell-whether-the-worlds-biggest-quantum-computer-is-a-quantum-computer/` 91

Kevin P. Murphy. A survey of pomdp solution techniques. *Environment*, 2, page X3, 2000. 35

Ajit Narayanan and Tammy Menneer. Quantum artificial neural network architectures and components. *Information Sciences*, 128(3), pages 231–255, 2000. DOI: 10.1016/s0020-0255(00)00055-4. 66

Makoto Naruse, Naoya Tate, Masashi Aono, and Motoichi Ohtsu. Information physics fundamentals of nanophotonics. *Reports on Progress in Physics*, 76(5), page 056401, 2013. DOI: 10.1088/0034-4885/76/5/056401. 33

Makoto Naruse, Martin Berthel, Aurélien Drezet, Serge Huant, Masashi Aono, Hirokazu Hori, and Song-Ju Kim. Single-photon decision maker. *Scientific Reports*, 5, 2015. DOI: 10.1038/srep13253. 44, 45

Rodion Neigovzen, Jorge L. Neves, Rudolf Sollacher, and Steffen J. Glaser. Quantum pattern recognition with liquid-state nuclear magnetic resonance. *Physical Review A*, 79(4), page 042321, 2009. DOI: 10.1103/physreva.79.042321. 65

Hartmut Neven, Vasil S. Denchev, Geordie Rose, and William G. Macready. Training a binary classifier with the quantum adiabatic algorithm, 2008. 58, 71

Hartmut Neven, Vasil S. Denchev, Geordie Rose, and William G. Macready. Training a large scale classifier with the quantum adiabatic algorithm, 2009. 19

Michael A. Nielsen and Isaac L. Chuang. *Quantum Computation and Quantum Information*. Cambridge University Press, New York, NY, 10th Anniversary ed., 2010a. DOI: 10.1017/CBO9780511976667.

Michael A. Nielsen and Isaac L. Chuang. Quantum information theory. In *Quantum Computation and Quantum Information*, pages 528–607, Cambridge University Press (CUP), 2010b. DOI: 10.1017/cbo9780511976667.016.

Tim Oates, Matthew D. Schmill, and Paul R. Cohen. A method for clustering the experiences of a mobile robot that accords with human judgments. In *Proc. of the 17th National Conference on Artificial Intelligence and 12th Conference on Innovative Applications of Artificial Intelligence*, pages 846–851, AAAI Press, 2000. http://dl.acm.org/citation.cfm?id=647288.721117 56

Bryan O'Gorman, R. Babbush, A. Perdomo-Ortiz, Alan Aspuru-Guzik, and Vadim Smelyanskiy. Bayesian network structure learning using quantum annealing. *The European Physical Journal Special Topics*, 224(1), pages 163–188, 2015. DOI: 10.1140/epjst/e2015-02349-9. 60, 62

Joni Pajarinen and Ville Kyrki. Robotic manipulation of multiple objects as a pomdp. *Artificial Intelligence*, 2015. DOI: 10.1016/j.artint.2015.04.001. 33

Christos H. Papadimitriou and John N. Tsitsiklis. The complexity of Markov decision processes. *Mathematics of Operations Research*, 12(3), pages 441–450, 1987. DOI: 10.1287/moor.12.3.441. 36

Giuseppe Davide Paparo, Vedran Dunjko, Adi Makmal, Miguel Angel Martin-Delgado, and Hans J. Briegel. Quantum speedup for active learning agents. *Physical Review X*, 4(3), page 031002, 2014. DOI: 10.1103/physrevx.4.031002. 33, 42, 43

Kalyanapuram R. Parthasarathy. *An Introduction to Quantum Stochastic Calculus*. Springer Science and Business Media, 2012. DOI: 10.1007/978-3-0348-0566-7. 87

Thomas Pellizzari, Simon A. Gardiner, Juan Ignacio Cirac, and Peter Zoller. Decoherence, continuous observation, and quantum computing: A cavity qed model. *Physical Review Letters*, 75(21), 1995. DOI: 10.1103/PhysRevLett.75.3788. 93

Raphael Pelossof, Andrew Miller, Peter Allen, and Tony Jebara. An svm learning approach to robotic grasping. In *Robotics and Automation, Proc. (ICRA'04), International Conference on*, volume 4, pages 3512–3518, IEEE, 2004. DOI: 10.1109/robot.2004.1308797. 56, 59

Joelle Pineau and Sebastian Thrun. High-level robot behavior control using pomdps. In *AAAI-02 Workshop on Cognitive Robotics*, volume 107, 2002. 33

Giovanni Pini, Arne Brutschy, Gianpiero Francesca, Marco Dorigo, and Mauro Birattari. Multi-armed bandit formulation of the task partitioning problem in swarm robotics. In *Lecture Notes in Computer Science*, pages 109–120. Springer, 2012. DOI: 10.1007/978-3-642-32650-9_10. 44

Dean A. Pomerleau. *Neural Network Perception for Mobile Robot Guidance*. Kluwer Academic Publishers, Norwell, MA, 1993. DOI: 10.1007/978-1-4615-3192-0. 62

Andrzej Pronobis, O. Martinez Mozos, and Barbara Caputo. Svm-based discriminative accumulation scheme for place recognition. In *Robotics and Automation, (ICRA) International Conference on*, pages 522–529, IEEE, 2008. DOI: 10.1109/robot.2008.4543260. 56, 59

Kristen L. Pudenz, Tameem Albash, and Daniel A. Lidar. Error-corrected quantum annealing with hundreds of qubits. *Nature Communications*, 5, 2014. DOI: 10.1038/ncomms4243. 19

Hong Qiao, Peng Zhang, Bo Zhang, and Suiwu Zheng. Learning an intrinsic-variable preserving manifold for dynamic visual tracking. *Systems, Man, and Cybernetics, Part B: Cybernetics, IEEE Transactions on*, 40(3), pages 868–880, 2010. DOI: 10.1109/tsmcb.2009.2031559. 70

Lawrence R. Rabiner and Biing-Hwang Juang. An introduction to hidden Markov models. *ASSP Magazine, IEEE*, 3(1), pages 4–16, 1986. DOI: 10.1109/massp.1986.1165342. 78

Timothy C. Ralph. Howard wiseman and gerard milburn: Quantum measurement and control. *Quantum Information Processing*, 11(1), pages 313–315, 2011. ISSN 1573-1332. DOI: 10.1007/s11128-011-0277-3. 79

Pramila Rani, Changchun Liu, Nilanjan Sarkar, and Eric Vanman. An empirical study of machine learning techniques for affect recognition in human-robot interaction. *Pattern Analysis and Applications*, 9(1), pages 58–69, 2006. DOI: 10.1007/s10044-006-0025-y. 47

Nathan Ratliff, David Bradley, J. Andrew Bagnell, and Joel Chestnutt. Boosting structured prediction for imitation learning. *Robotics Institute*, page 54, 2007. 70, 72

Patrick Rebentrost, Masoud Mohseni, and Seth Lloyd. Quantum support vector machine for big data classification. *Physical Review Letters*, 113(13), page 130503, 2014. DOI: 10.1103/physrevlett.113.130503. 58, 59

Eleanor Rieffel and Wolfgang Polak. An introduction to quantum computing for non-physicists. *ACM Computing Surveys*, 32(3), pages 300–335, 2000. DOI: 10.1145/367701.367709. 13

Eleanor G. Rieffel, Davide Venturelli, Bryan O'Gorman, Minh B. Do, Elicia M. Prystay, and Vadim N. Smelyanskiy. A case study in programming a quantum annealer for hard operational planning problems. *Quantum Information Processing*, 14(1), pages 1–36, 2015. DOI: 10.1007/s11128-014-0892-x. 25, 28, 30

Ronald L. Rivest, Adi Shamir, and Len Adleman. A method for obtaining digital signatures and public-key cryptosystems. *Communications of the ACM*, 21(2), pages 120–126, 1978. DOI: 10.1145/359340.359342. 20

Troels F. Rønnow, Zhihui Wang, Joshua Job, Sergio Boixo, Sergei V. Isakov, David Wecker, John M. Martinis, Daniel A. Lidar, and Matthias Troyer. Defining and detecting quantum speedup. *Science*, 345(6195), pages 420–424, 2014. DOI: 10.1126/science.1252319. 18, 96

Frank Rosenblatt. The perceptron: A probabilistic model for information storage and organization in the brain. *Psychological Review*, 65(6), page 386, 1958. DOI: 10.1037/h0042519. 62

Axel Rottmann, Óscar Martínez Mozos, Cyrill Stachniss, and Wolfram Burgard. Semantic place classification of indoor environments with mobile robots using boosting. In *Proc. of the 20th National Conference on Artificial Intelligence*, Volume 3, pages 1306–1311, AAAI Press, 2005. http://dl.acm.org/citation.cfm?id=1619499.1619543 70, 72

Nicholas Roy and Geoffrey J. Gordon. Exponential family pca for belief compression in pomdps. In S. Becker, S. Thrun, and K. Obermayer, Eds., *Advances in Neural Information Processing Systems 15*, pages 1667–1674, MIT Press, 2003. http://papers.nips.cc/paper/2319-exponential-family-pca-for-belief-compression-in-pomdps.pdf 51

Nicholas Roy, Geoffrey J. Gordon, and Sebastian Thrun. Finding approximate pomdp solutions through belief compression. *J. Artif. Intell. Res. (JAIR)*, 23, pages 1–40, 2005. DOI: doi:10.1613/jair.1496. 35

David E. Rumelhart, Geoffrey E. Hinton, and Ronald J. Williams. Learning internal representations by error propagation. In *Readings in Cognitive Science*, pages 399–421, Elsevier BV, 1988. DOI: 10.1016/b978-1-4832-1446-7.50035-2. 63

László Ruppert. Towards Kalman filtering of finite quantum systems, 2012. 82, 83

Stuart J. Russell and Peter Norvig. *Artificial Intelligence: A Modern Approach*. Pearson Education, 2nd ed., 2003. 27, 35

Ruslan Salakhutdinov and Geoffrey E Hinton. Deep Boltzmann machines. In *International Conference on Artificial Intelligence and Statistics*, pages 448–455, 2009. DOI: 10.1007/978-1-4899-7502-7_31-1. 64

Lawrence K. Saul and Sam T. Roweis. Think globally, fit locally: Unsupervised learning of low dimensional manifolds. *The Journal of Machine Learning Research*, 4, pages 119–155, 2003. 69

Tobias Schaetz, Chris R. Monroe, and Tilman Esslinger. Focus on quantum simulation. *New Journal of Physics*, 15(8), page 085009, 2013. DOI: 10.1088/1367-2630/15/8/085009. 43

Maria Schuld, Ilya Sinayskiy, and Francesco Petruccione. The quest for a quantum neural network. *Quantum Information Processing*, 13(11), pages 2567–2586, 2014. DOI: 10.1007/s11128-014-0809-8. 67

Maria Schuld, Ilya Sinayskiy, and Francesco Petruccione. An introduction to quantum machine learning. *Contemporary Physics*, 56(2), pages 172–185, 2015. DOI: 10.1080/00107514.2014.964942. 47, 48

Hadayat Seddiqi and Travis S. Humble. Adiabatic quantum optimization for associative memory recall. *Frontiers in Physics*, 2, 2014. DOI: 10.3389/fphy.2014.00079. 65

Burr Settles. Active learning literature survey, 52(55–66), page 11, University of Wisconsin, Madison, 2010. 33

Guy Shani, Joelle Pineau, and Robert Kaplow. A survey of point-based pomdp solvers. *Autonomous Agents and Multi-Agent Systems*, 27(1), pages 1–51, 2013. DOI: 10.1007/s10458-012-9200-2. 35

Vivek V. Shende, Stephen S. Bullock, and Igor L. Markov. Synthesis of quantum-logic circuits. *IEEE Transactions on Computer-Aided Design of Integrated Circuits and Systems*, 25(6), pages 1000–1010, 2006. DOI: 10.1109/TCAD.2005.855930. 97

Peter W. Shor. Polynomial-time algorithms for prime factorization and discrete logarithms on a quantum computer. *SIAM Review*, 41(2), pages 303–332, 1999. DOI: 10.1137/s0036144598347011. 20

Bailu Si, Michael J. Herrmann, and Klaus Pawelzik. Gain-based exploration: From multi-armed bandits to partially observable environments. In *Natural Computation, (ICNC) 3rd International Conference on*, volume 1, pages 177–182, IEEE, 2007. DOI: 10.1109/icnc.2007.395. 44

Bruno Siciliano, Lorenzo Sciavicco, Luigi Villani, and Giuseppe Oriolo. *Robotics: Modeling, Planning and Control*. Springer Science and Business Media, 2010. 83, 87

Reid Simmons and Sven Koenig. Probabilistic robot navigation in partially observable environments. In *IJCAI*, volume 95, pages 1080–1087, 1995. 33

Vadim N. Smelyanskiy, Eleanor G. Rieffel, Sergey I. Knysh, Colin P. Williams, Mark W. Johnson, Murray C. Thom, William G. Macready, and Kristen L. Pudenz. A near-term quantum computing approach for hard computational problems in space exploration, 2012. 25, 28

Robert S. Smith, Michael J. Curtis, and William J. Zeng. A practical quantum instruction set architecture. https://arxiv.org/abs/1608.03355 https://arxiv.org/abs/1608.03355 97

Paul Smolensky. Information processing in dynamical systems: Foundations of harmony theory, 1986. 64

John A. Smolin and Graeme Smith. Classical signature of quantum annealing. *Frontiers in Physics*, 2, 2013. DOI: 10.3389/fphy.2014.00052. 94, 95

Kathy-Anne Brickman Soderberg, Nathan Gemelke, and Cheng Chin. Ultracold molecules: Vehicles to scalable quantum information processing. *New Journal of Physics*, 11(5), 2009. DOI: 10.1088/1367-2630/11/5/055022. 93

Dan Song, Carl Henrik Ek, Kai Huebner, and Danica Kragic. Multivariate discretization for Bayesian network structure learning in robot grasping. In *Robotics and Automation (ICRA), International Conference on*, pages 1944–1950, IEEE, 2011. DOI: 10.1109/icra.2011.5979666. 60, 62

Matthijs T. J. Spaan and Nikos Vlassis. A point-based pomdp algorithm for robot planning. In *Robotics and Automation, Proc. (ICRA'04) International Conference on*, volume 3, pages 2399–2404, IEEE, 2004. DOI: 10.1109/robot.2004.1307420. 33

Vishnu Srivastava, Paul Reverdy, and Naomi Ehrich Leonard. On optimal foraging and multi-armed bandits. In *Communication, Control, and Computing (Allerton), 51st Annual Allerton Conference on*, pages 494–499, IEEE, 2013. DOI: 10.1109/allerton.2013.6736565. 44

Andrew Steane. The ion trap quantum information processor. *Applied Physics B: Lasers and Optics*, 64(6), pages 623–643, 1997a. DOI: 10.1007/s003400050225. 92

Andrew Steane. The ion trap quantum information processor. *Applied Physics B: Lasers and Optics*, 64(6), pages 623–643, 1997b. DOI: 10.1007/s003400050225. 92

Andrew Steane. Quantum computing. *Reports on Progress in Physics*, 61(2), page 117, 1998. DOI: 10.1088/0034-4885/61/2/002. 13

Richard S. Sutton and Andrew G. Barto. *Reinforcement learning: An Introduction*, volume 1. MIT press Cambridge, 1998. DOI: 10.1109/tnn.1998.712192. 33

Nikhil Swaminathan. Why does the brain need so much power? *Scientific American*, 29(04), page 2998, 2008. 10

Ryan Sweke, Ilya Sinayskiy, and Francesco Petruccione. Simulation of single-qubit open quantum systems. *Physical Review A*, 90(2), page 022331, 2014. DOI: 10.1103/physreva.90.022331. 79

Mario Szegedy.   Quantum speed-up of Markov chain based algorithms.   In *Foundations of Computer Science, Proc. of the 45th Annual IEEE Symposium on*, pages 32–41, 2004. DOI: 10.1109/focs.2004.53. 42

Luís Tarrataca and Andreas Wichert. Tree search and quantum computation. *Quantum Information Processing*, 10(4), pages 475–500, 2011. DOI: 10.1007/s11128-010-0212-z. 25, 27, 28

Joshua B. Tenenbaum, Vin De Silva, and John C. Langford.   A global geometric framework for nonlinear dimensionality reduction.   *Science*, 290(5500), pages 2319–2323, 2000. DOI: 10.1126/science.290.5500.2319. 68

Laura Kathrine Wehde Thomsen, Stefano Mancini, and Howard Mark Wiseman. Spin squeezing via quantum feedback. *Physical Review A*, 65(6), page 061801, 2002. DOI: 10.1103/physreva.65.061801. 86

Sebastian Thrun, Dieter Fox, Wolfram Burgard, and Frank Dellaert.   Robust Monte Carlo localization for mobile robots.   *Artificial Intelligence*, 128(1), pages 99–141, 2001. DOI: 10.1016/s0004-3702(01)00069-8. 47

Géza Tóth, Craig S. Lent, P. Douglas Tougaw, Yuriy Brazhnik, Weiwen Weng, Wolfgang Porod, Ruey-Wen Liu, and Yih-Fang Huang. Quantum cellular neural networks, 2000. DOI: 10.1006/spmi.1996.0104. 93

Carlo A. Trugenberger.   Probabilistic quantum memories.   *Physical Review Letters*, 87(6), page 067901, 2001. DOI: 10.1103/physrevlett.87.067901. 65

Ramon van Handel, John K. Stockton, and Hideo Mabuchi.   Modelling and feedback control design for quantum state preparation. *Journal of Optics B: Quantum and Semiclassical Optics*, 7(10), page S179, 2005. DOI: 10.1088/1464-4266/7/10/001. 86

Dan Ventura and Tony Martinez.   Quantum associative memory.   *Information Sciences*, 124(1), pages 273–296, 2000. DOI: 10.1016/S0020-0255(99)00101-2. 64, 65

Vandi Verma, Geoff Gordon, Reid Simmons, and Sebastian Thrun.   Real-time fault diagnosis [robot fault diagnosis]. *Robotics amd Automation Magazine, IEEE*, 11(2), pages 56–66, 2004. DOI: 10.1109/mra.2004.1310942. 78

Mathukumalli Vidyasagar.   The realization problem for hidden Markov models: The complete realization problem.   In *Decision and Control, European Control Conference, (CDC-ECC'05) 44th IEEE Conference on*, pages 6632–6637, 2005. DOI: 10.1109/cdc.2005.1583227. 79

Nikos Vlassis and Ben Krose. Robot environment modeling via principal component regression. In *Intelligent Robots and Systems, (IROS'99), Proc. of the International Conference on*, volume 2, pages 677–682, IEEE, 1999. DOI: 10.1109/iros.1999.812758. 51

Alexis De Vos. *Reversible Computing: Fundamentals, Quantum Computing, and Applications.* Wiley-VCH, 2010. DOI: 10.1002/9783527633999. 9, 12

Jeffrey M. Walls and Ryan W. Wolcott. Learning loop closures by boosting, 2011. 70, 72

Guoming Wang. Quantum algorithms for curve fitting, 2014. 53

Lei Wang, Troels F. Rønnow, Sergio Boixo, Sergei V. Isakov, Zhihui Wang, David Wecker, Daniel A. Lidar, John M. Martinis, and Matthias Troyer. Comment on: Classical signature of quantum annealing, 2013. 95

Yazhen Wang. Quantum computation and quantum information. *Statistical Science*, 27(3), pages 373–394, 2012. DOI: 10.1214/11-STS378.

Kilian Q. Weinberger, Fei Sha, and Lawrence K. Saul. Learning a kernel matrix for nonlinear dimensionality reduction. In *Proc. of the 21st International Conference on Machine Learning*, page 106, ACM, 2004. DOI: 10.1145/1015330.1015345. 69

Nathan Wiebe, Daniel Braun, and Seth Lloyd. Quantum algorithm for data fitting. *Physical Review Letters*, 109(5), page 050505, 2012. DOI: 10.1103/physrevlett.109.050505. 53

Nathan Wiebe, Ashish Kapoor, and Krysta M. Svore. Quantum deep learning, 2014. 66

Peter Wittek. *Quantum Machine Learning: What Quantum Computing Means to Data Mining.* Elsevier Insights, Elsevier Science, 2014. https://books.google.com/books?id=92hzA wAAQBAJ 47, 67

Denis F. Wolf, Gaurav S. Sukhatme, Dieter Fox, and Wolfram Burgard. Autonomous terrain mapping and classification using hidden Markov models. In *Robotics and Automation, (ICRA) Proc. of the International Conference on*, pages 2026–2031, IEEE, 2005. DOI: 10.1109/robot.2005.1570411. 78

Zang Xizhe, Zhao Jie, Wang Chen, and Cai Hegao. Study on a svm-based data fusion method. In *Robotics, Automation and Mechatronics, Conference on*, volume 1, pages 413–415, IEEE, 2004. DOI: 10.1109/ramech.2004.1438955. 56, 59

Mingsheng Ying, Yangjia Li, Nengkun Yu, and Yuan Feng. Model-checking linear-time properties of quantum systems. *ACM Transactions on Computational Logic*, 15(3), pages 22:1–22:31, 2014. DOI: 10.1145/2629680. 87

Shenggang Ying and Mingsheng Ying. Reachability analysis of quantum Markov decision processes, 2014. 33, 36, 37, 87

Michail Zak and Colin P. Williams. Quantum neural nets. *International Journal of Theoretical Physics*, 37(2), pages 651–684, 1998. DOI: 10.1023/A:1026656110699. 66

Jing Zhang, Yu-xi Liu, and Franco Nori. Cooling and squeezing the fluctuations of a nanome-chanical beam by indirect quantum feedback control. *Physical Review A*, 79(5), page 052102, 2009. DOI: 10.1103/physreva.79.052102. 87

Kaiguang Zhao, Sorin Popescu, Xuelian Meng, Yong Pang, and Muge Agca. Characterizing forest canopy structure with lidar composite metrics and machine learning. *Remote Sensing of Environment*, 115(8), pages 1978–1996, 2011. DOI: 10.1016/j.rse.2011.04.001. 47

Hongjun Zhou and Shigeyuki Sakane. Mobile robot localization using active sensing based on Bayesian network inference. *Robotics and Autonomous Systems*, 55(4), pages 292–305, 2007. DOI: 10.1016/j.robot.2006.11.005. 60, 62

Authors' Biographies

PRATEEK TANDON

Prateek Tandon is the founder of the Quantum Robotics Group. He obtained his Ph.D. from the Robotics Institute at Carnegie Mellon University and his B.S. in Computer Engineering, Computer Science from the University of Southern California. His Ph.D. dissertation focused on Bayesian methods for sensor fusion and extracting structure from multiple noisy data observations. Having previously written on optimizing many classical machine learning methods for robotic sensing, improving fault tolerance and runtime scalability, Prateek's research interests include exploring the new possibilities that quantum methods and quantum machine learning hold for robotic perception, learning, and planning.

STANLEY LAM

Stanley Lam holds an M.Sc. in Social Research Methods from the London School of Economics and a B.S. with Honors in Business Administration (Global Management) from the University of Southern California. His work deploys research methodologies toward solving real-world problems in business and technology. Some prior work includes modeling Silicon Valley labor patterns using Gini coefficients, graphing public sector workforces with nonparametric estimators, and developing user studies for assistive technology. He led discussions in the Quantum Robotics Group on recent hardware advances in quantum research, and is dedicated to understanding the latest quantum computing implementations, their prospects for scalability and commercialization, and the challenges which remain in order for quantum robotics to be fully realized.

BENJAMIN SHIH

Benjamin Shih is a Ph.D. student in Mechanical and Aerospace Engineering at the University of California, San Diego. He holds an M.S. and B.S. in Electrical and Computer Engineering from Carnegie Mellon University (2013) and has been a research assistant at École Polytechnique Fédérale de Lausanne. His main area of research is soft robotics, an intersection of control, materials, and cognitive science. Ben led discussion of quantum control in the Quantum Robotics Group.

TANAY MEHTA

Tanay Mehta is a Ph.D. student in Computer Science at Northeastern University. He holds a B.S. in mathematics from the University of Southern California. His research is focused on the theoretical aspects of computer science and the fundamental limits of computation. He is interested in the role of pseudo-randomness in complexity theory and cryptography. Tanay served as the lead for quantum information theory discussions in the Quantum Robotics Group.

ALEX MITEV

Alex Mitev holds a Ph.D. in Electrical Engineering from the University of Arizona. His main research interests are in statistical modeling, model order reduction, machine learning, and dynamic programming. He has published several papers analyzing problems of timing analysis, macro-modeling, and the impact of process variations for VLSI circuits. Alex led discussions of quantum circuits in the Quantum Robotics Group.

ZHIYANG ONG

Zhiyang Ong is an Electrical Engineering Ph.D. student at Texas A&M University. He received his M.S. in Electrical Engineering from the University of Southern California and his Bachelor of Engineering (Electrical and Electronic) from the University of Adelaide. He has interned at the Institute of Microelectronics and Bioinformatics Institute in Singapore, Symantec Corporation in Santa Monica, California, and the University of Trento in Trento, Italy. He has also studied at the University of Verona and National Taiwan University. He is a student member of IEEE and ACM and is a member of IEEE-Eta Kappa Nu. Zhiyang led discussions of quantum modeling checking in the Quantum Robotics Group.

Index